FREE Test Taking Tips DVD Offer

To help us better serve you, we have developed a Test Taking Tips DVD that we would like to give you for FREE. **This DVD covers world-class test taking tips that you can use to be even more successful when you are taking your test.**

All that we ask is that you email us your feedback about your study guide. Please let us know what you thought about it – whether that is good, bad or indifferent.

To get your **FREE Test Taking Tips DVD**, email freedvd@studyguideteam.com with "FREE DVD" in the subject line and the following information in the body of the email:

a. The title of your study guide.

b. Your product rating on a scale of 1-5, with 5 being the highest rating.

c. Your feedback about the study guide. What did you think of it?

d. Your full name and shipping address to send your free DVD.

If you have any questions or concerns, please don't hesitate to contact us at freedvd@studyguideteam.com.

Thanks again!

ATI TEAS 6 Science Test Study Guide 2019 & 2020

ATI TEAS Science Study Manual & Two Practice Tests

Test Prep Books

Table of Contents

Quick Overview 1
Test-Taking Strategies 2
FREE DVD OFFER 6
Introduction to the ATI TEAS 7
Human Anatomy and Physiology 9
- Anatomy and Physiology of a Human 9
- Respiratory System 18
- Cardiovascular System 21
- Gastrointestinal System 23
- Neuromuscular System 26
- Reproductive System 31
- Integumentary System 33
- Endocrine System 34
- Genitourinary System 36
- Immune System 38
- Skeletal System 41

Life and Physical Sciences 44
- Basic Macromolecules in a Biological System 44
- Chromosomes, Genes, and DNA 50
- Mendel's Laws of Heredity 67
- Basic Atomic Structure 79
- Characteristic Properties of Substances 85
- Changes in States of Matter 90
- Chemical Reactions 92

Scientific Reasoning 99
- Scientific Measurement Using Laboratory Tools 99
- Critiquing a Scientific Explanation Using Logic and Evidence 109
- Relationships Among Events, Objects, and Processes 113
- Design of a Scientific Investigation 129

Practice Test #1 135
Answer Explanations #1 145
Practice Test #2 153
Answer Explanations #2 161

Quick Overview

As you draw closer to taking your exam, effective preparation becomes more and more important. Thankfully, you have this study guide to help you get ready. Use this guide to help keep your studying on track and refer to it often.

This study guide contains several key sections that will help you be successful on your exam. The guide contains tips for what you should do the night before and the day of the test. Also included are test-taking tips. Knowing the right information is not always enough. Many well-prepared test takers struggle with exams. These tips will help equip you to accurately read, assess, and answer test questions.

A large part of the guide is devoted to showing you what content to expect on the exam and to helping you better understand that content. In this guide are practice test questions so that you can see how well you have grasped the content. Then, answer explanations are provided so that you can understand why you missed certain questions.

Don't try to cram the night before you take your exam. This is not a wise strategy for a few reasons. First, your retention of the information will be low. Your time would be better used by reviewing information you already know rather than trying to learn a lot of new information. Second, you will likely become stressed as you try to gain a large amount of knowledge in a short amount of time. Third, you will be depriving yourself of sleep. So be sure to go to bed at a reasonable time the night before. Being well-rested helps you focus and remain calm.

Be sure to eat a substantial breakfast the morning of the exam. If you are taking the exam in the afternoon, be sure to have a good lunch as well. Being hungry is distracting and can make it difficult to focus. You have hopefully spent lots of time preparing for the exam. Don't let an empty stomach get in the way of success!

When travelling to the testing center, leave earlier than needed. That way, you have a buffer in case you experience any delays. This will help you remain calm and will keep you from missing your appointment time at the testing center.

Be sure to pace yourself during the exam. Don't try to rush through the exam. There is no need to risk performing poorly on the exam just so you can leave the testing center early. Allow yourself to use all of the allotted time if needed.

Remain positive while taking the exam even if you feel like you are performing poorly. Thinking about the content you should have mastered will not help you perform better on the exam.

Once the exam is complete, take some time to relax. Even if you feel that you need to take the exam again, you will be well served by some down time before you begin studying again. It's often easier to convince yourself to study if you know that it will come with a reward!

Test-Taking Strategies

1. Predicting the Answer

When you feel confident in your preparation for a multiple-choice test, try predicting the answer before reading the answer choices. This is especially useful on questions that test objective factual knowledge. By predicting the answer before reading the available choices, you eliminate the possibility that you will be distracted or led astray by an incorrect answer choice. You will feel more confident in your selection if you read the question, predict the answer, and then find your prediction among the answer choices. After using this strategy, be sure to still read all of the answer choices carefully and completely. If you feel unprepared, you should not attempt to predict the answers. This would be a waste of time and an opportunity for your mind to wander in the wrong direction.

2. Reading the Whole Question

Too often, test takers scan a multiple-choice question, recognize a few familiar words, and immediately jump to the answer choices. Test authors are aware of this common impatience, and they will sometimes prey upon it. For instance, a test author might subtly turn the question into a negative, or he or she might redirect the focus of the question right at the end. The only way to avoid falling into these traps is to read the entirety of the question carefully before reading the answer choices.

3. Looking for Wrong Answers

Long and complicated multiple-choice questions can be intimidating. One way to simplify a difficult multiple-choice question is to eliminate all of the answer choices that are clearly wrong. In most sets of answers, there will be at least one selection that can be dismissed right away. If the test is administered on paper, the test taker could draw a line through it to indicate that it may be ignored; otherwise, the test taker will have to perform this operation mentally or on scratch paper. In either case, once the obviously incorrect answers have been eliminated, the remaining choices may be considered. Sometimes identifying the clearly wrong answers will give the test taker some information about the correct answer. For instance, if one of the remaining answer choices is a direct opposite of one of the eliminated answer choices, it may well be the correct answer. The opposite of obviously wrong is obviously right! Of course, this is not always the case. Some answers are obviously incorrect simply because they are irrelevant to the question being asked. Still, identifying and eliminating some incorrect answer choices is a good way to simplify a multiple-choice question.

4. Don't Overanalyze

Anxious test takers often overanalyze questions. When you are nervous, your brain will often run wild, causing you to make associations and discover clues that don't actually exist. If you feel that this may be a problem for you, do whatever you can to slow down during the test. Try taking a deep breath or counting to ten. As you read and consider the question, restrict yourself to the particular words used by the author. Avoid thought tangents about what the author *really* meant, or what he or she was *trying* to say. The only things that matter on a multiple-choice test are the words that are actually in the question. You must avoid reading too much into a multiple-choice question, or supposing that the writer meant something other than what he or she wrote.

5. No Need for Panic

It is wise to learn as many strategies as possible before taking a multiple-choice test, but it is likely that you will come across a few questions for which you simply don't know the answer. In this situation, avoid panicking. Because most multiple-choice tests include dozens of questions, the relative value of a single wrong answer is small. As much as possible, you should compartmentalize each question on a multiple-choice test. In other words, you should not allow your feelings about one question to affect your success on the others. When you find a question that you either don't understand or don't know how to answer, just take a deep breath and do your best. Read the entire question slowly and carefully. Try rephrasing the question a couple of different ways. Then, read all of the answer choices carefully. After eliminating obviously wrong answers, make a selection and move on to the next question.

6. Confusing Answer Choices

When working on a difficult multiple-choice question, there may be a tendency to focus on the answer choices that are the easiest to understand. Many people, whether consciously or not, gravitate to the answer choices that require the least concentration, knowledge, and memory. This is a mistake. When you come across an answer choice that is confusing, you should give it extra attention. A question might be confusing because you do not know the subject matter to which it refers. If this is the case, don't eliminate the answer before you have affirmatively settled on another. When you come across an answer choice of this type, set it aside as you look at the remaining choices. If you can confidently assert that one of the other choices is correct, you can leave the confusing answer aside. Otherwise, you will need to take a moment to try to better understand the confusing answer choice. Rephrasing is one way to tease out the sense of a confusing answer choice.

7. Your First Instinct

Many people struggle with multiple-choice tests because they overthink the questions. If you have studied sufficiently for the test, you should be prepared to trust your first instinct once you have carefully and completely read the question and all of the answer choices. There is a great deal of research suggesting that the mind can come to the correct conclusion very quickly once it has obtained all of the relevant information. At times, it may seem to you as if your intuition is working faster even than your reasoning mind. This may in fact be true. The knowledge you obtain while studying may be retrieved from your subconscious before you have a chance to work out the associations that support it. Verify your instinct by working out the reasons that it should be trusted.

8. Key Words

Many test takers struggle with multiple-choice questions because they have poor reading comprehension skills. Quickly reading and understanding a multiple-choice question requires a mixture of skill and experience. To help with this, try jotting down a few key words and phrases on a piece of scrap paper. Doing this concentrates the process of reading and forces the mind to weigh the relative importance of the question's parts. In selecting words and phrases to write down, the test taker thinks about the question more deeply and carefully. This is especially true for multiple-choice questions that are preceded by a long prompt.

9. Subtle Negatives

One of the oldest tricks in the multiple-choice test writer's book is to subtly reverse the meaning of a question with a word like *not* or *except*. If you are not paying attention to each word in the question, you can easily be led astray by this trick. For instance, a common question format is, "Which of the following is...?" Obviously, if the question instead is, "Which of the following is not...?," then the answer will be quite different. Even worse, the test makers are aware of the potential for this mistake and will include one answer choice that would be correct if the question were not negated or reversed. A test taker who misses the reversal will find what he or she believes to be a correct answer and will be so confident that he or she will fail to reread the question and discover the original error. The only way to avoid this is to practice a wide variety of multiple-choice questions and to pay close attention to each and every word.

10. Reading Every Answer Choice

It may seem obvious, but you should always read every one of the answer choices! Too many test takers fall into the habit of scanning the question and assuming that they understand the question because they recognize a few key words. From there, they pick the first answer choice that answers the question they believe they have read. Test takers who read all of the answer choices might discover that one of the latter answer choices is actually *more* correct. Moreover, reading all of the answer choices can remind you of facts related to the question that can help you arrive at the correct answer. Sometimes, a misstatement or incorrect detail in one of the latter answer choices will trigger your memory of the subject and will enable you to find the right answer. Failing to read all of the answer choices is like not reading all of the items on a restaurant menu: you might miss out on the perfect choice.

11. Spot the Hedges

One of the keys to success on multiple-choice tests is paying close attention to every word. This is never truer than with words like almost, most, some, and sometimes. These words are called "hedges" because they indicate that a statement is not totally true or not true in every place and time. An absolute statement will contain no hedges, but in many subjects, the answers are not always straightforward or absolute. There are always exceptions to the rules in these subjects. For this reason, you should favor those multiple-choice questions that contain hedging language. The presence of qualifying words indicates that the author is taking special care with his or her words, which is certainly important when composing the right answer. After all, there are many ways to be wrong, but there is only one way to be right! For this reason, it is wise to avoid answers that are absolute when taking a multiple-choice test. An absolute answer is one that says things are either all one way or all another. They often include words like *every, always, best*, and *never*. If you are taking a multiple-choice test in a subject that doesn't lend itself to absolute answers, be on your guard if you see any of these words.

12. Long Answers

In many subject areas, the answers are not simple. As already mentioned, the right answer often requires hedges. Another common feature of the answers to a complex or subjective question are qualifying clauses, which are groups of words that subtly modify the meaning of the sentence. If the question or answer choice describes a rule to which there are exceptions or the subject matter is complicated, ambiguous, or confusing, the correct answer will require many words in order to be expressed clearly and accurately. In essence, you should not be deterred by answer choices that seem excessively long. Oftentimes, the author of the text will not be able to write the correct answer without offering some qualifications and modifications. Your job is to read the answer choices thoroughly and completely and to select the one that most accurately and precisely answers the question.

13. Restating to Understand

Sometimes, a question on a multiple-choice test is difficult not because of what it asks but because of how it is written. If this is the case, restate the question or answer choice in different words. This process serves a couple of important purposes. First, it forces you to concentrate on the core of the question. In order to rephrase the question accurately, you have to understand it well. Rephrasing the question will concentrate your mind on the key words and ideas. Second, it will present the information to your mind in a fresh way. This process may trigger your memory and render some useful scrap of information picked up while studying.

14. True Statements

Sometimes an answer choice will be true in itself, but it does not answer the question. This is one of the main reasons why it is essential to read the question carefully and completely before proceeding to the answer choices. Too often, test takers skip ahead to the answer choices and look for true statements. Having found one of these, they are content to select it without reference to the question above. Obviously, this provides an easy way for test makers to play tricks. The savvy test taker will always read the entire question before turning to the answer choices. Then, having settled on a correct answer choice, he or she will refer to the original question and ensure that the selected answer is relevant. The mistake of choosing a correct-but-irrelevant answer choice is especially common on questions related to specific pieces of objective knowledge. A prepared test taker will have a wealth of factual knowledge at his or her disposal, and should not be careless in its application.

15. No Patterns

One of the more dangerous ideas that circulates about multiple-choice tests is that the correct answers tend to fall into patterns. These erroneous ideas range from a belief that B and C are the most common right answers, to the idea that an unprepared test-taker should answer "A-B-A-C-A-D-A-B-A." It cannot be emphasized enough that pattern-seeking of this type is exactly the WRONG way to approach a multiple-choice test. To begin with, it is highly unlikely that the test maker will plot the correct answers according to some predetermined pattern. The questions are scrambled and delivered in a random order. Furthermore, even if the test maker was following a pattern in the assignation of correct answers, there is no reason why the test taker would know which pattern he or she was using. Any attempt to discern a pattern in the answer choices is a waste of time and a distraction from the real work of taking the test. A test taker would be much better served by extra preparation before the test than by reliance on a pattern in the answers.

FREE DVD OFFER

Don't forget that doing well on your exam includes both understanding the test content and understanding how to use what you know to do well on the test. We offer a completely FREE Test Taking Tips DVD that covers world class test taking tips that you can use to be even more successful when you are taking your test.

All that we ask is that you email us your feedback about your study guide. To get your **FREE Test Taking Tips DVD**, email freedvd@studyguideteam.com with "FREE DVD" in the subject line and the following information in the body of the email:

- The title of your study guide.
- Your product rating on a scale of 1-5, with 5 being the highest rating.
- Your feedback about the study guide. What did you think of it?
- Your full name and shipping address to send your free DVD.

Introduction to the ATI TEAS

Background of the ATI TEAS

The Test of Essential Academic Skills (TEAS) is a standardized test created and distributed by Assessment Technologies Institute (ATI) to examine the test taker's aptitude for skill sets fundamental to a career in nursing. As such, the TEAS is used by nursing schools and allied health schools in the United States and Canada as a chief criterion for admission. The TEAS is currently in its sixth iteration, known as the ATI TEAS.

The ATI TEAS is a nationwide test, with no variation in the difficulty of content among the versions given from state to state; the content of the TEAS is a standard measure for entry-level skills and abilities for nursing applicants. However, the required minimum TEAS scores, like prerequisite courses or application distinctions, can vary widely between schools and programs. Because the TEAS is used for admission to nursing and allied health programs, the majority of TEAS takers are high school diploma or GED graduates pursuing a career in nursing or are applicants to programs requiring prerequisite academic coursework. These applicants can range in background from sophomore-level collegiate to professionals looking to change careers into a healthcare field.

Test Administration

The ATI TEAS may be administered by a nursing or allied health school or a PSI testing center. The testing schedule is chosen by each individual facility, and the regularity of testing can vary from major metropolitan areas (where it may be offered multiple times per week in several locations) to sparsely populated towns (where it may be offered once per month every hundred miles). The test will also be proctored to make certain that testing protocols are enforced. Test takers can register at atitesting.com or directly through the school to which they wish to apply, as most nursing schools offer the test on-campus periodically throughout the year. The cost to take the ATI TEAS is set by the local administrator, but it usually ranges from $83 to $105.

Students may retake the ATI TEAS, but most schools have limitations such as the number of days students must wait between attempts, or the number of attempts students may make in a given period. For instance, many schools will accept the higher score of two attempts within a twelve-month period, but will not consider a third attempt until twelve months has lapsed since the first attempt. Disability accommodations are generally available and made by contacting the local test administration site.

Test Format

The ATI TEAS is comprised of 170 multiple choice questions with four possible answer choices given for each question. The questions are divided between four subject areas—Reading, Mathematics, Science, and English and Language Usage.

Subject Area	**Questions**	**Time Limit (minutes)**
Reading	53	64
Mathematics	36	54
Science	53	63
English & Language Usage	28	28
Total	**170**	**209**

Test takers are given 209 minutes to take the ATI TEAS, broken down among the four subject areas. Once a test taker begins a subject area, the timer will start. When the timer expires, the test taker may stretch, go to the bathroom, or otherwise relax before the timer begins for the next section. The ATI TEAS is offered in both a pencil-and-paper version and a computerized version, depending on the preference of the local test administrator. ATI TEAS test takers are not permitted to use cell phones, but four-function calculators are permitted now on the ATI TEAS.

Scoring

Shortly after the examination, test takers will receive several different numbered scores with their ATI TEAS results, including scores on all the various sections and subsections. This includes the national and state-level percentile rankings that a test taker has achieved. This rank is equal to the percentage of test takers from nationwide samples that scored equal to or lower than the test taker. Higher percentiles indicate more correct answers and fewer people who scored higher than the given test taker. However, although this may be of interest to the test taker, schools do not use this score to base acceptance decisions on.

Instead, schools typically look at the Composite Individual Total Score. The composite score is a good determination of the performance of the test taker and is calculated by averaging the test taker's performance from each section of the test. The national average composite score generally lies between 65% correct and 75% correct, and many schools have a minimum required score in this range, but each school and program chooses their own minimum standards; more prestigious schools require higher minimum scores and vice versa. Besides meeting minimum requirements, TEAS scores can play a part in the acceptance process as one of the factors like previous GPAs and extracurricular activities that demonstrate the competitiveness of the applicant, but this too varies from school to school.

Recent Developments

ATI TEAS is the sixth version of the TEAS, and is offered starting August 31, 2016. Before August 31, 2016, the fifth version of the TEAS (known as the TEAS V) was the current version of the TEAS. The ATI TEAS is similar in difficulty, but with slight differences in the amount of emphasis placed on various subjects, such as the reduction of the English & Language Usage section and the elimination of the Earth Science subsection. Four-function calculators are now permitted on the ATI TEAS.

Human Anatomy and Physiology

Anatomy and Physiology of a Human

Anatomy may be defined as the structural makeup of an organism. The study of anatomy may be divided into microscopic/fine anatomy and macroscopic/gross anatomy. Fine anatomy concerns itself with viewing the features of the body with the aid of a microscope, while gross anatomy concerns itself with viewing the features of the body with the naked eye. **Physiology** refers to the functions of an organism and it examines the chemical or physical functions that help the body function appropriately.

Levels of Organization of the Human Body

All the parts of the human body are built of individual units called **cells.** Groups of similar cells are arranged into **tissues,** different tissues are arranged into **organs**, and organs working together form entire **organ systems**. The human body has twelve organ systems that govern circulation, digestion, immunity, hormones, movement, support, coordination, urination & excretion, reproduction (male and female), respiration, and general protection.

Here are some of the systems of the body:

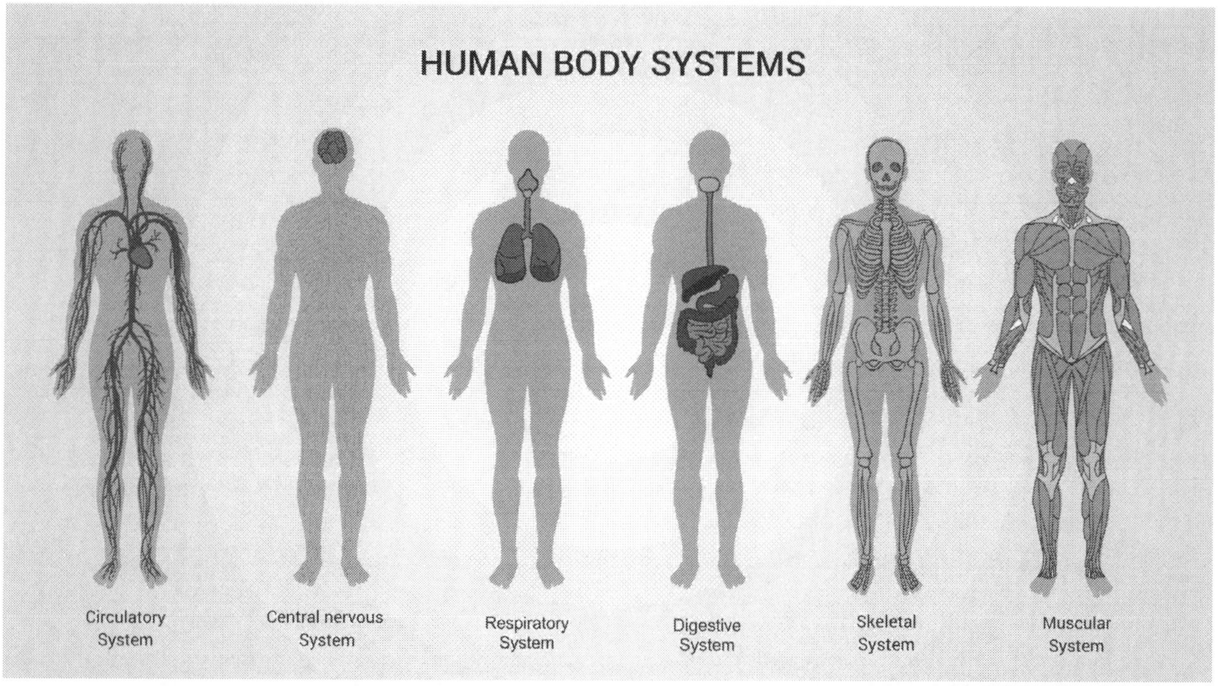

Specialization happens within organs and organ systems, which allow the cells in that organ to turn off genes and processes not related to that function, thereby conserving energy. For example, the digestive system carries out the process of breaking down food so that nutrients can be taken up by cells. The mouth chews the food and creates saliva with the enzymes necessary for the breakdown of some starches. Swallowing brings the food into the esophagus and down to the stomach, where acidic gastric juices are secreted that break down food further and kill any bacteria present in the food. From the stomach, food travels to the small intestines, where the food is broken down into basic molecular units and absorbed through the walls of the intestines into the blood stream. Waste continues on through the large intestines and is eliminated from the body as feces.

Organs are organized into organ systems that carry out biological functions. These organ systems must communicate and work together through interactions for optimal function. For example, the urinary-tract system is an organ system that comprises several organs, including the kidney and the bladder. The role of the urinary-tract system is to filter waste and extra fluid from the blood and remove it from the body. The kidneys and the bladder have to work together to complete this process. Throughout the day, the kidneys receive blood and then filter waste products and excess fluid from the blood. The waste produced by the kidneys is urine. As urine is produced, it travels through the ureter and into the bladder, which acts as a holding system until the person is ready to urinate.

As another example of organ systems working together, muscle contraction is a voluntary process that requires the nervous system and the musculoskeletal system to function cooperatively. When a person wants to move a muscle, an action potential is sent through the axon of a motor nerve cell, called a motor neuron. Between the motor neuron and the muscle it innervates, there is a gap, or synapse. When the action potential reaches the end of the motor neuron, the neuron releases a neurotransmitter, called acetylcholine, into the synapse. This chemical then signals the muscle and creates an action potential in the muscle, which causes the muscle to contract.

Body Cavities

The body is partitioned into different hollow spaces that house organs. The human body contains the following cavities:

- Cranial cavity: The cranial cavity is surrounded by the skull and contains organs such as the brain and pituitary gland.
- Thoracic cavity: The thoracic cavity is encircled by the sternum (breastbone) and ribs. It contains organs such as the lungs, heart, trachea (windpipe), esophagus, and bronchial tubes.
- Abdominal cavity: The abdominal cavity is separated from the thoracic cavity by the diaphragm. It contains organs such as the stomach, gallbladder, liver, small intestines, and large intestines. The abdominal organs are held in place by a membrane called the peritoneum.
- Pelvic cavity: The pelvic cavity is enclosed by the pelvis, or bones of the hip. It contains organs such as the urinary bladder, urethra, ureters, anus, and rectum. It contains the reproductive organs as well. In females, the pelvic cavity also contains the uterus.
- Spinal cavity: The spinal cavity is surrounded by the vertebral column. The vertebral column has five regions: cervical, thoracic, lumbar, sacral, and coccygeal. The spinal cord runs through the middle of the spinal cavity.

Human Tissues

Human tissues can be grouped into four categories:

1. **Muscle:** Muscle tissue supports the body and allows it to move, and muscle cells (called **fibers**) are special in their ability to contract. There are three distinct types of muscle tissue: skeletal, smooth, and cardiac. Skeletal muscle is voluntary, or under conscious control, and is usually attached to bones. Most body movement is directly caused by the contraction of skeletal muscle. Smooth muscle is typically involuntary, or not under conscious control, and is found in blood vessels, the walls of hollow organs, and the urinary bladder. Cardiac muscle is involuntary and found in the heart, where it helps propel blood throughout the body.

2. **Nervous:** Nervous tissue is unique in that it is able to coordinate information from sensory organs as well as communicate the proper behavioral responses. Neurons, or nerve cells, are the workhorses of the nervous system. They communicate via action potentials (electrical signals) and neurotransmitters (chemical signals).

3. **Epithelial:** Epithelial tissue covers the external surfaces of organs and lines many of the body's cavities. Epithelial tissue helps to protect the body from invasion by microbes (bacteria, viruses, parasites), fluid loss, and injury.

Epithelial cell shapes can be:

1. Squamous: cells with a flat shape
2. Cuboidal: cells with a cubed shape
3. Columnar: cells shaped like a column

Epithelial cells can be arranged in four patterns:

1. Simple: a type of epithelium composed solely from a single layer of cells
2. Stratified: a type of epithelium composed of multiple layers of cells
3. Pseudostratified: appears stratified but actually consists of only one layer of cells
4. Transitional: a type of epithelium noted for its ability to expand and contract

4. **Connective:** Connective tissue supports and connects the tissues and organs of the body. Connective tissue is composed of cells dispersed throughout a matrix which can be gel, liquid, protein fibers, or salts. The primary protein fibers in the matrix are collagen (for strength), elastin (for flexibility), and reticulum (for support). Connective tissue can be categorized as either loose or dense. Examples of connective tissue include bones, cartilage, ligaments, tendons, blood, and adipose (fat) tissue.

Three Primary Body Planes

A plane is an imaginary flat surface. The three primary planes of the human body are frontal, sagittal, and transverse. The frontal, or coronal, plane is a vertical plane that divides the body or organ into front (anterior) and back (posterior) portions. The sagittal, or lateral, plane is a vertical plane divides the body or organ into right and left sides. The transverse, or axial, plane is a horizontal plane that divides the body or organ into upper and lower portions. In medical imaging, computed tomography (CT) scans are oriented only in the transverse plane; while magnetic resonance imaging (MRI) scans may be oriented in any of the three planes.

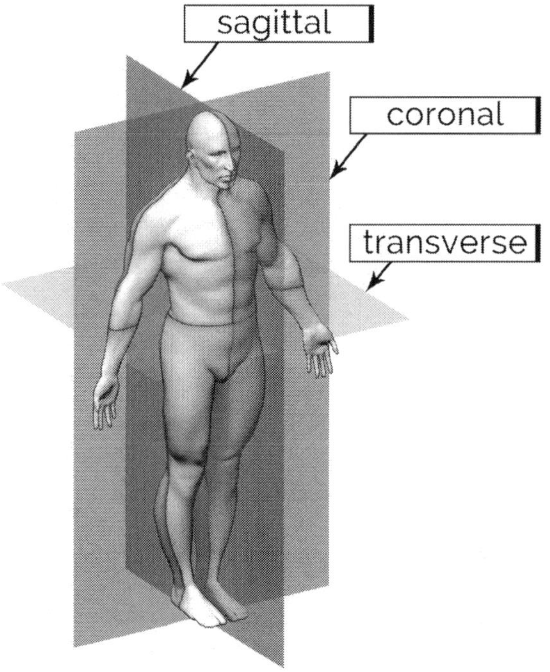

Terms of Direction

Medial refers to a structure being closer to the midline of the body. For example, the nose is medial to the eyes.

Lateral refers to a structure being farther from the midline of the body, and it is the opposite of medial. For example, the eyes are lateral to the nose.

Proximal refers to a structure or body part located near an attachment point. For example, the elbow is proximal to the wrist.

Distal refers to a structure or body part located far from an attachment point, and it is the opposite of proximal. For example, the wrist is distal to the elbow.

Anterior means toward the front in humans. For example, the lips are anterior to the teeth. The term **ventral** can be used in place of anterior.

Posterior means toward the back in humans, and it is the opposite of anterior. For example, the teeth are posterior to the lips. The term **dorsal** can be used in place of posterior.

Superior means above and refers to a structure closer to the head. For example, the head is superior to the neck. The terms **cephalic** or **cranial** may be used in place of superior.

Inferior means below and refers to a structure farther from the head, and it is the opposite of superior. For example, the neck is inferior to the head. The term **caudal** may be used in place of inferior.

Superficial refers to a structure closer to the surface. For example, the muscles are superficial because they are just beneath the surface of the skin.

Deep refers to a structure farther from the surface, and it is the opposite of superficial. For example, the femur is a deep structure lying beneath the muscles.

Body Regions

Terms for general locations on the body include:

- Cervical: relating to the neck
- Clavicular: relating to the clavicle, or collarbone
- Ocular: relating to the eyes
- Acromial: relating to the shoulder
- Cubital: relating to the elbow
- Brachial: relating to the arm
- Carpal: relating to the wrist
- Thoracic : relating to the chest
- Abdominal: relating to the abdomen
- Pubic: relating to the groin
- Pelvic: relating to the pelvis, or bones of the hip
- Femoral: relating to the femur, or thigh bone
- Geniculate: relating to the knee
- Pedal: relating to the foot
- Palmar: relating to the palm of the hand
- Plantar: relating to the sole of the foot

Abdominopelvic Regions and Quadrants

The abdominopelvic region may be defined as the combination of the abdominal and the pelvic cavities. The region's upper border is the breasts and its lower border is the groin region. The region is divided into the following nine sections:

- Right hypochondriac: region below the cartilage of the ribs
- Epigastric: region above the stomach between the hypochondriac regions
- Left hypochondriac: region below the cartilage of the ribs
- Right lumbar: region of the waist
- Umbilical: region between the lumbar regions where the umbilicus (navel) is located
- Left lumbar: region of the waist
- Right inguinal: region of the groin
- Hypogastric: region below the stomach between the inguinal regions
- Left inguinal: region of the groin

A simpler way to describe the abdominopelvic area would be to divide it into the following quadrants:

- Right upper quadrant (RUQ): Encompasses the right hypochondriac, right lumbar, epigastric, and umbilical regions.
- Right lower quadrant (RLQ): Encompasses the right lumbar, right inguinal, hypogastric, and umbilical regions.
- Left upper quadrant (LUQ): Encompasses the left hypochondriac, left lumbar, epigastric, and umbilical regions.
- Left lower quadrant (LLQ): Encompasses the left lumbar, left inguinal, hypogastric, and umbilical regions.

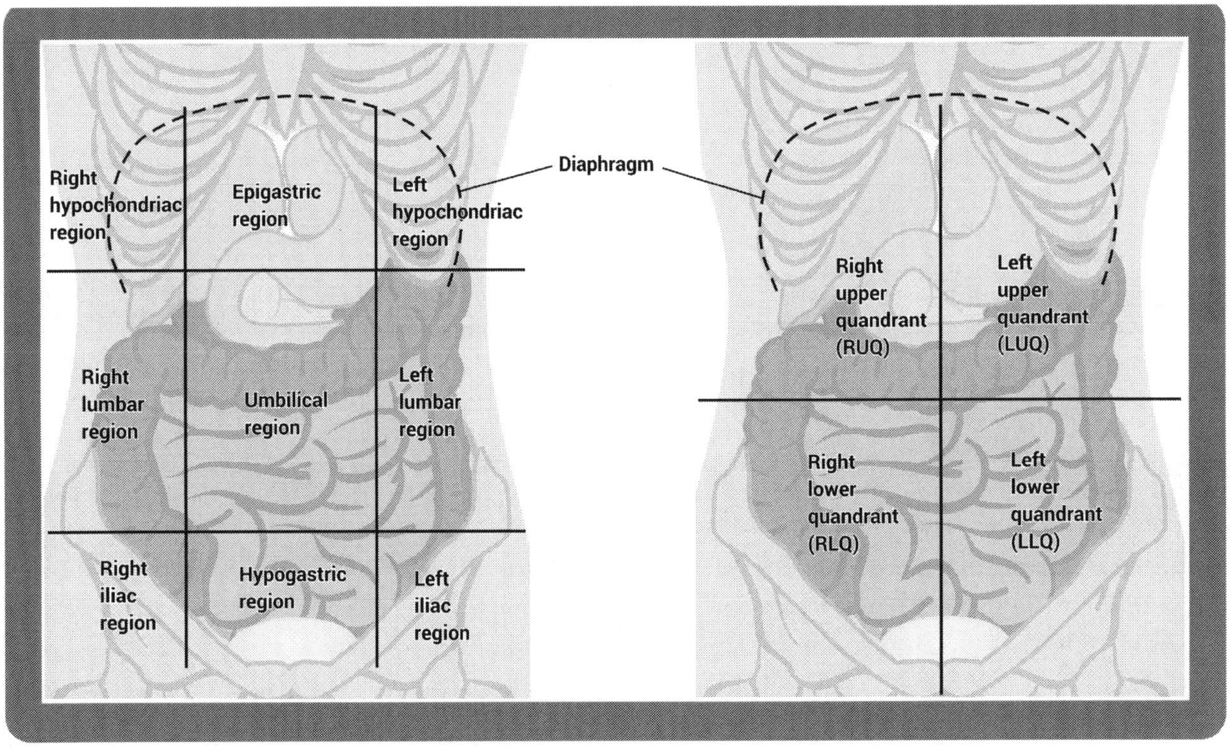

Cell Structure and Function

The cell is the main functional and structural component of all living organisms. Robert Hooke, an English scientist, coined the term "cell" in 1665. Hooke's discovery laid the groundwork for the **cell theory**, which is composed of three principles:

1. All organisms are composed of cells.
2. All existing cells are created from other living cells.
3. The cell is the most fundamental unit of life.

Organisms can be unicellular (composed of one cell) or multicellular (composed of many cells). All cells must be bounded by a cell membrane, be filled with cytoplasm of some sort, and be coded by a genetic sequence.

The cell membrane separates a cell's internal and external environments. It is a selectively permeable membrane, which usually only allows the passage of certain molecules by diffusion. Phospholipids and proteins are crucial components of all cell membranes. The cytoplasm is the cell's internal environment and is aqueous, or water-based. The genome represents the genetic material inside the cell that is passed on from generation to generation.

Prokaryotes and Eukaryotes

Prokaryotic cells are much smaller than eukaryotic cells. The majority of prokaryotes are unicellular, while the majority of eukaryotes are multicellular. Prokaryotic cells have no nucleus, and their genome is found in an area known as the **nucleoid.** They also do not have membrane-bound organelles, which are "little organs" that perform specific functions within a cell.

Eukaryotic cells have a proper nucleus containing the genome. They also have numerous membrane-bound organelles such as lysosomes, endoplasmic reticula (rough and smooth), Golgi complexes, and mitochondria. The majority of prokaryotic cells have cell walls, while most eukaryotic cells (aside from plant and fungal cells) do not have cell walls. The DNA of prokaryotic cells is contained in a single circular chromosome, while the DNA of eukaryotic cells is contained in multiple linear chromosomes. Prokaryotic cells divide using binary fission, while eukaryotic cells divide using mitosis. Examples of prokaryotes are bacteria and archaea while examples of eukaryotes are animals and plants.

Nuclear Parts of a Cell

Nucleus (plural nuclei): Houses a cell's genetic material, deoxyribonucleic acid (DNA), which is used to form chromosomes. A single nucleus is the defining characteristic of eukaryotic cells. The nucleus of a cell controls gene expression. It ensures genetic material is transmitted from one generation to the next.

Chromosomes: Complex thread-like arrangements composed of DNA that is found in a cell's nucleus. Humans have 23 pairs of chromosomes for a total of 46.

Chromatin: An aggregate of genetic material consisting of DNA and proteins that forms chromosomes during cell division.

Nucleolus (plural: nucleoli): The largest component of the nucleus of a eukaryotic cell. With no membrane, the primary function of the nucleolus is the production of ribosomes, which are crucial to the synthesis of proteins.

Cell Membranes

Cell membranes encircle the cell's cytoplasm, separating the intracellular environment from the extracellular environment. They are selectively permeable, which enables them to control molecular traffic entering and exiting cells. Cell membranes are made of a double layer of phospholipids studded with proteins. Cholesterol is also dispersed in the phospholipid bilayer of cell membranes to provide stability. The proteins in the phospholipid bilayer aid the transport of molecules across cell membranes.

Scientists use the term "fluid mosaic model" to refer to the arrangement of phospholipids and proteins in cell membranes. In that model, phospholipids have a head region and a tail region. The head region of the phospholipids is attracted to water (hydrophilic), while the tail region is repelled by it (hydrophobic). Because they are hydrophilic, the heads of the phospholipids are facing the water, pointing inside and outside of the cell. Because they are hydrophobic, the tails of the phospholipids are oriented inward between both head regions. This orientation constructs the phospholipid bilayer.

Cell membranes have the distinct trait of selective permeability. The fact that cell membranes are **amphiphilic** (having hydrophilic and hydrophobic zones) contributes to this trait. As a result, cell membranes are able to regulate the flow of molecules in and out of the cell.

Factors relating to molecules such as size, polarity, and solubility determine their likelihood of passage across cell membranes. Small molecules are able to diffuse easily across cell membranes compared to large molecules. Polarity refers to the charge present in a molecule. Polar molecules have regions, or poles, of positive and negative charge and are water soluble, while nonpolar molecules have no charge and are fat-soluble. Solubility refers to the ability of a substance, called a solute, to dissolve in a solvent. A soluble substance can be dissolved in a solvent, while an insoluble substance cannot be dissolved in a solvent. Nonpolar, fat-soluble substances have a much easier time passing through cell membranes compared to polar, water-soluble substances.

<u>Passive Transport Mechanisms</u>

Passive transport refers to the migration of molecules across a cell membrane that does not require energy. The three types of passive transport include simple diffusion, facilitated diffusion, and osmosis.

Simple diffusion relies on a concentration gradient, or differing quantities of molecules inside or outside of a cell. During simple diffusion, molecules move from an area of high concentration to an area of low concentration. Facilitated diffusion utilizes carrier proteins to transport molecules across a cell membrane. Osmosis refers to the transport of water across a selectively permeable membrane. During osmosis, water moves from a region of low solute concentration to a region of high solute concentration.

<u>Active Transport Mechanisms</u>

Active transport refers to the migration of molecules across a cell membrane that requires energy. It's a useful way to move molecules from an area of low concentration to an area of high concentration. Adenosine triphosphate (ATP), the currency of cellular energy, is needed to work against the concentration gradient.

Active transport can involve carrier proteins that cross the cell membrane to pump molecules and ions across the membrane, like in facilitated diffusion. The difference is that active transport uses the energy from ATP to drive this transport, as typically the ions or molecules are going against their concentration gradient. For example, glucose pumps in the kidney pump all of the glucose into the cells from the lumen of the nephron even though there is a higher concentration of glucose in the cell than in the lumen. This is because glucose is a precious food source and the body wants to conserve as much as possible. Pumps can either send a molecule in one direction, multiple molecules in the same direction (**symports**), or multiple molecules in different directions (**antiports**).

Active transport can also involve the movement of membrane-bound particles, either into a cell (**endocytosis**) or out of a cell (**exocytosis**). The three major forms of endocytosis are: **pinocytosis**, where the cell is drinking and intakes only small molecules; **phagocytosis**, where the cell is eating and intakes large particles or small organisms; and receptor-mediated endocytosis, where the cell's membrane splits off to form an internal vesicle as a response to molecules activating receptors on its surface. Exocytosis is the inverse of endocytosis, and the membranes of the vesicle join to that of the cell's surface while the molecules inside the vesicle are released outside. This is common in nervous and muscle tissue for the release of neurotransmitters and in endocrine cells for the release of hormones. The two major categories of exocytosis are excretion and secretion. **Excretion** is defined as the removal of waste from a cell. **Secretion** is defined as the transport of molecules from a cell such as hormones or enzymes.

Structure and Function of Cellular Organelles

Organelles are specialized structures that perform specific tasks in a cell. The term literally means "little organ." Most organelles are membrane bound and serve as sites for the production or degradation of chemicals. The following are organelles found in eukaryotic cells:

Nucleus: A cell's nucleus contains genetic information in the form of DNA. The nucleus is surrounded by the nuclear envelope. A single nucleus is the defining characteristic of eukaryotic cells. The nucleus is also the most important organelle of the cell. It contains the nucleolus, which manufactures ribosomes (another organelle) that are crucial in protein synthesis (also called gene expression).

Mitochondria: Mitochondria are oval-shaped and have a double membrane. The inner membrane has multiple folds called cristae. Mitochondria are responsible for the production of a cell's energy in the form of adenosine triphosphate (ATP). ATP is the principal energy transfer molecule in eukaryotic cells. Mitochondria also participate in cellular respiration.

Rough Endoplasmic Reticulum: The rough endoplasmic reticulum (RER) is composed of linked membranous sacs called cisternae with ribosomes attached to their external surfaces. The RER is responsible for the production of proteins that will eventually get shipped out of the cell.

Smooth Endoplasmic Reticulum: The smooth endoplasmic reticulum (SER) is composed of linked membranous sacs called cisternae without ribosomes, which distinguishes it from the RER. The SER's main function is the production of carbohydrates and lipids that can be created expressly for the cell or to modify the proteins from the RER that will eventually get shipped out of the cell.

Golgi Apparatus: The Golgi apparatus is located next to the SER. Its main function is the final modification, storage, and shipping of products (proteins, carbohydrates, and lipids) from the endoplasmic reticulum.

Lysosomes: Lysosomes are specialized vesicles that contain enzymes capable of digesting food, surplus organelles, and foreign invaders such as bacteria and viruses. They often destroy dead cells in order to recycle cellular components. Lysosomes are found only in animal cells.

Secretory Vesicles: Secretory vesicles transport and deliver molecules into or out of the cell via the cell membrane. Endocytosis refers to the movement of molecules into a cell via secretory vesicles. Exocytosis refers to the movement of molecules out of a cell via secretory vesicles.

Ribosomes: Ribosomes are not membrane bound. They are responsible for the production of proteins as specified from DNA instructions. Ribosomes can be free or bound.

Cilia and Flagella: Cilia are specialized hair-like projections on some eukaryotic cells that aid in movement, while flagella are long, whip-like projections that are used in the same capacity.

Here is an illustration of the cell:

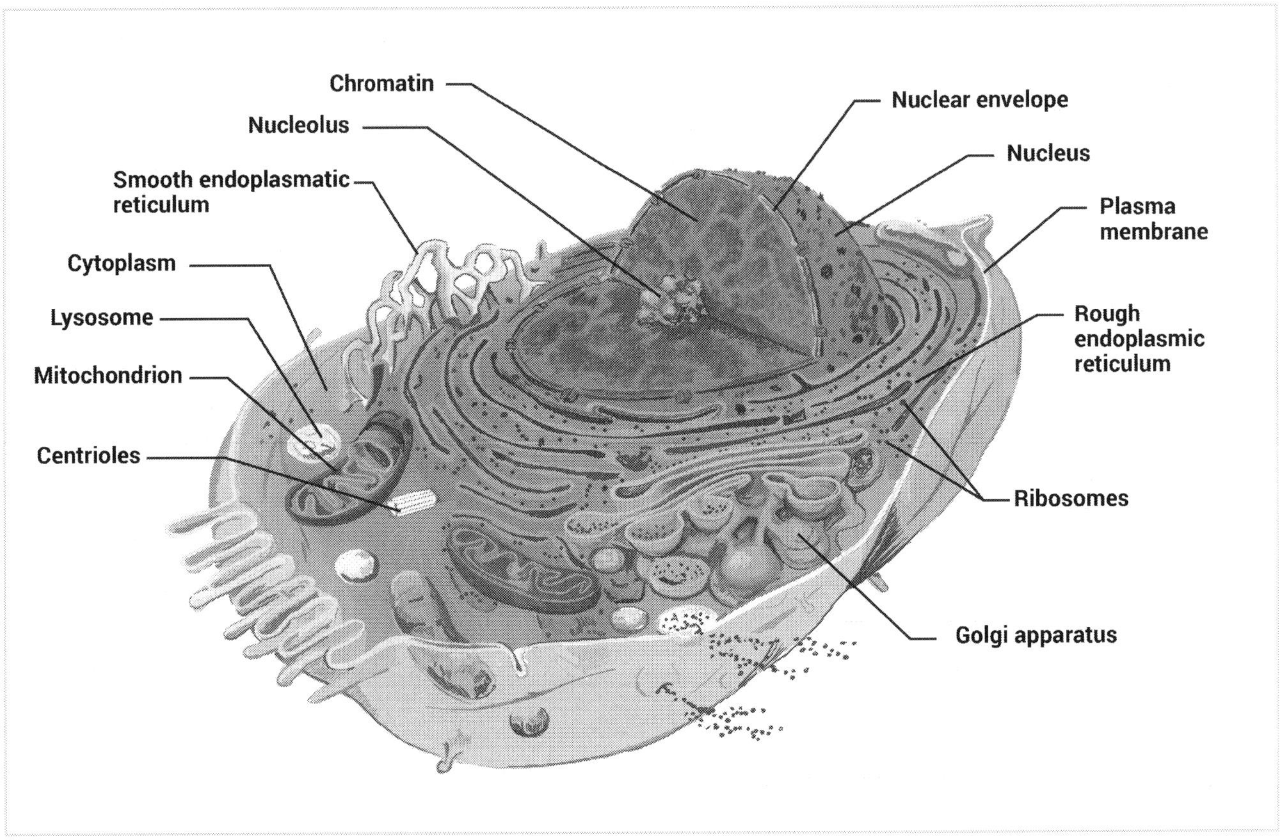

The following organelles are not found in animal cells:

Cell Walls: Cell walls can be found in plants, bacteria, and fungi, and are made of cellulose, peptidoglycan, and lignin according to the organism it is found around. Each of these materials is a type of sugar recognized as a structural carbohydrate. The carbohydrates: are rigid structures located outside of the cell membrane. Cell walls function to protect the cell, help maintain a cell's shape, and provide structural support.

Vacuoles: Plant cells have central vacuoles, which are essentially a membrane surrounding a body of water. They may store nutrients or waste products. Since vacuoles are large, they also help to support the structure of plant cells.

Chloroplasts: Chloroplasts are membrane-bound organelles that perform photosynthesis. They contain structural units called thylakoids. Chlorophyll, a green pigment that circulates within the thylakoids, harnesses light energy (sunlight) and helps convert it into chemical energy (glucose).

Respiratory System

The respiratory system enables breathing and supports the energy-making process in our cells. The respiratory system transports an essential reactant, oxygen, to cells so that they can produce energy in their mitochondria via cellular respiration. The respiratory system also removes carbon dioxide, a waste product of cellular respiration.

This system is divided into the upper respiratory system and the lower respiratory system. The upper system comprises the nose, the nasal cavity and sinuses, and the pharynx. The lower respiratory system comprises the larynx (voice box), the trachea (windpipe), the small passageways leading to the lungs, and the lungs.

The pathway of oxygen to the bloodstream begins with the nose and the mouth. Upon inhalation, air enters the nose and mouth and passes into the sinuses where it gets warmed, filtered, and humidified. The throat, or the pharynx, allows the entry of both food and air; however, only air moves into the trachea, or windpipe, since the epiglottis covers the trachea during swallowing and prevents food from entering. The trachea contains mucus and cilia. The mucus traps many airborne pathogens while the cilia act as bristles that sweep the pathogens away toward the top of the trachea where they are either swallowed or coughed out.

The trachea itself has two vocal cords at the top which make up the larynx. At its bottom, the trachea forks into two major bronchi—one for each lung. These bronchi continue to branch into smaller and smaller bronchioles before terminating in grape-like air sacs called alveoli; these alveoli are surrounded by capillaries and provide the body with an enormous amount of surface area to exchange oxygen and carbon dioxide gases, in a process called external respiration.

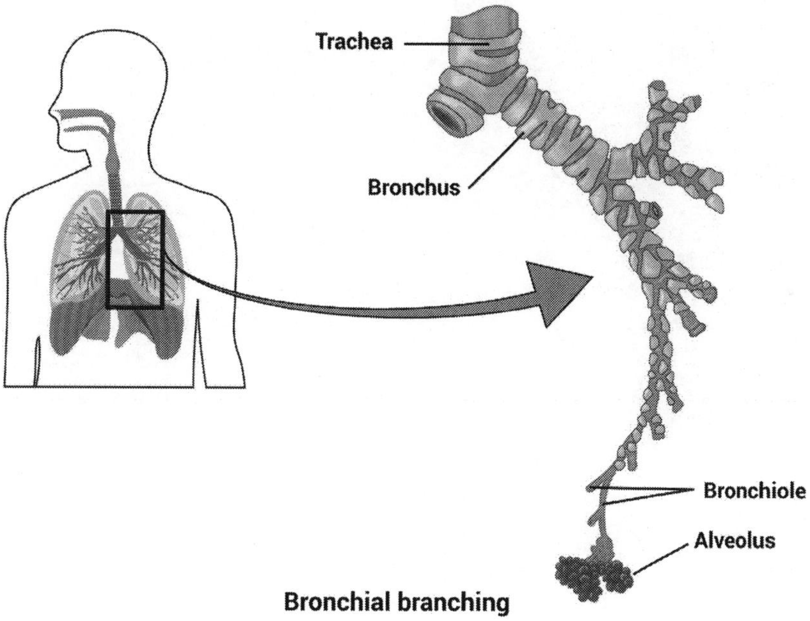

Bronchial branching

In total, the lungs contain about 1500 miles of airway passages. The right lung is divided into three lobes (superior, middle, and inferior), and the left lung is divided into two lobes (superior and inferior).

The left lung is smaller than the right lung, likely because it shares its space in the chest cavity with the heart.

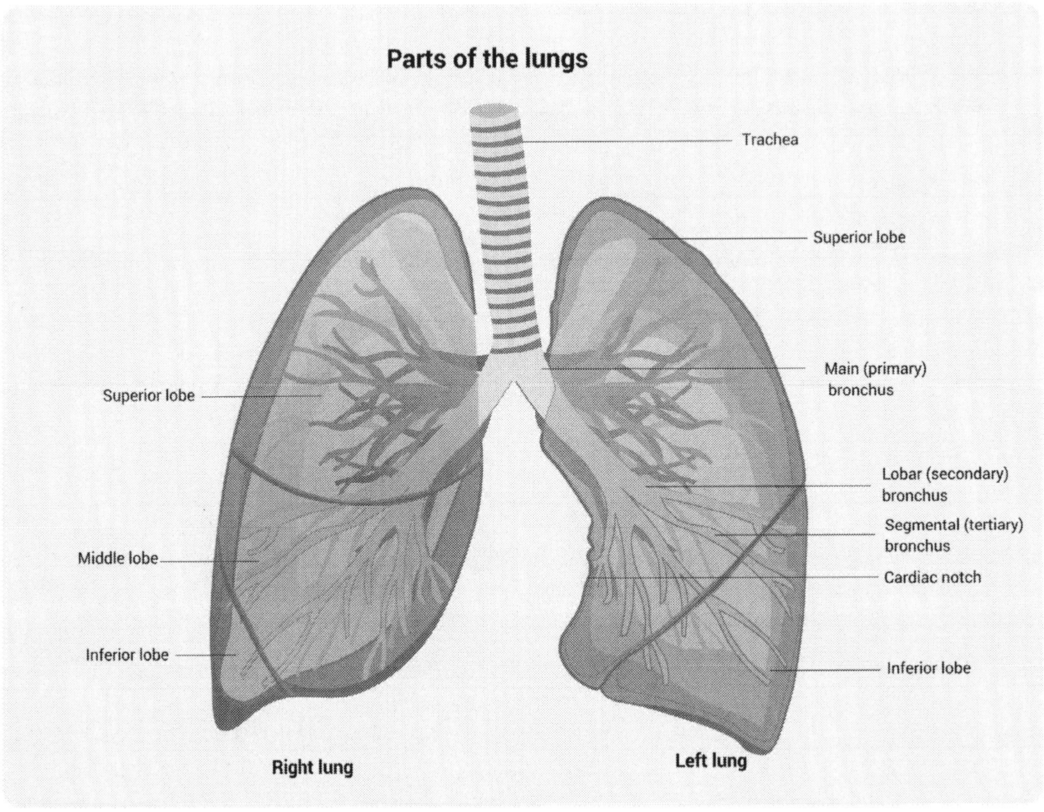

A flat muscle underneath the lungs called the diaphragm controls breathing. When the diaphragm contracts, the volume of the chest cavity increases and indirectly decreases its air pressure. This decrease in air pressure creates a vacuum, and the lungs pull in air to fill the space. This difference in air pressure that pulls the air from outside of the body into the lungs in a process called negative pressure breathing.

Upon inhalation or inspiration, oxygen in the alveoli diffuses into the capillaries to be carried by blood to cells throughout the body, in a process called internal respiration. A protein called hemoglobin in red blood cells easily bonds with oxygen, removing it from the blood and allowing more oxygen to diffuse in. This protein allows the blood to take in 60 times more oxygen than the body could without it, and this explains how oxygen can become so concentrated in blood even though it is only 21% of the atmosphere. While oxygen diffuses from the alveoli into the capillaries, carbon dioxide diffuses from the capillaries into the alveoli. When the diaphragm relaxes, the elastic lungs snap back to their original shape; this decreases the volume of the chest cavity and increases the air pressure until it is back to normal. This increased air pressure pushes the carbon dioxide waste from the alveoli through exhalation or **expiration**.

The autonomic nervous system controls breathing. The medulla oblongata gets feedback regarding the carbon dioxide levels in the blood, and will send a message to the diaphragm that it is time for a contraction. While breathing can be voluntary, it is mostly under autonomic control.

Functions of the Respiratory System

The respiratory system has many functions. Most importantly, it provides a large area for gas exchange between the air and the circulating blood. It protects the delicate respiratory surfaces from environmental variations and defends them against pathogens. It is responsible for producing the sounds that the body makes for speaking and singing, as well as for non-verbal communication. It also helps regulate blood volume and blood pressure by releasing vasopressin, and it is a regulator of blood pH due to its control over carbon dioxide release, as the aqueous form of carbon dioxide is the chief buffering agent in blood.

Cardiovascular System

The cardiovascular system (also called the circulatory system) is a network of organs and tubes that transport blood, hormones, nutrients, oxygen, and other gases to cells and tissues throughout the body. The major components of the circulatory system are the blood vessels, blood, and heart.

Blood Vessels

In the circulatory system, blood vessels are responsible for transporting blood throughout the body. The three major types of blood vessels in the circulatory system are arteries, veins, and capillaries. Arteries carry blood from the heart to the rest of the body. Veins carry blood from the body to the heart. Capillaries connect arteries to veins and form networks that exchange materials between the blood and the cells.

In general, arteries are stronger and thicker than veins, as they withstand high pressures exerted by the blood as the heart pumps it through the body. Arteries control blood flow through either **vasoconstriction** (narrowing of the blood vessel's diameter) or **vasodilation** (widening of the blood vessel's diameter). The blood in veins is under much lower pressures, so veins have valves to prevent the backflow of blood.

Most of the exchange between the blood and tissues takes place through the capillaries. There are three types of capillaries: continuous, fenestrated, and sinusoidal.

Continuous capillaries are made up of epithelial cells tightly connected together. As a result, they limit the types of materials that pass into and out of the blood. Continuous capillaries are the most common type of capillary. Fenestrated capillaries have openings that allow materials to be freely exchanged between the blood and tissues. They are commonly found in the digestive, endocrine, and urinary systems. Sinusoidal capillaries have larger openings and allow proteins and blood cells through. They are found primarily in the liver, bone marrow, and spleen.

Blood

Blood is vital to the human body. It is a liquid connective tissue that serves as a transport system for supplying cells with nutrients and carrying away their wastes. The average adult human has five to six quarts of blood circulating through their body. Approximately 55% of blood is plasma (the fluid portion), and the remaining 45% is composed of solid cells and cell parts.

There are three major types of blood cells:

- Red blood cells, or erythrocytes, transport oxygen throughout the body. They contain a protein called hemoglobin that allows them to carry oxygen. The iron in the hemoglobin gives the cells and the blood their red colors.
- White blood cells, or leukocytes, are responsible for fighting infectious diseases and maintaining the immune system. There are five types of white blood cells: neutrophils, lymphocytes, eosinophils, monocytes, and basophils.
- Platelets are cell fragments which play a central role in the blood clotting process.

All blood cells in adults are produced in the bone marrow—red blood cells from red marrow and white blood cells from yellow marrow.

Heart

The heart is a two-part, muscular pump that forcefully pushes blood throughout the human body. The human heart has four chambers—two upper atria and two lower ventricles, a pair on the left and a pair on the right. Anatomically, *left* and *right* correspond to the sides of the body that the patient themselves would refer to as left and right.

Four valves help to section off the chambers from one another. Between the right atrium and ventricle, the three flaps of the **tricuspid valve** keep blood from backflowing from the ventricle to the atrium, similar to how the two flaps of the **mitral valve** work between the left atrium and ventricle. As these two valves lie between an atrium and a ventricle, they are referred to as **atrioventricular (AV) valves**. The other two valves are **semilunar (SL)** and control blood flow into the two great arteries leaving the ventricles. The **pulmonary valve** connects the right ventricle to the pulmonary artery while the **aortic valve** connects the left ventricle to the aorta.

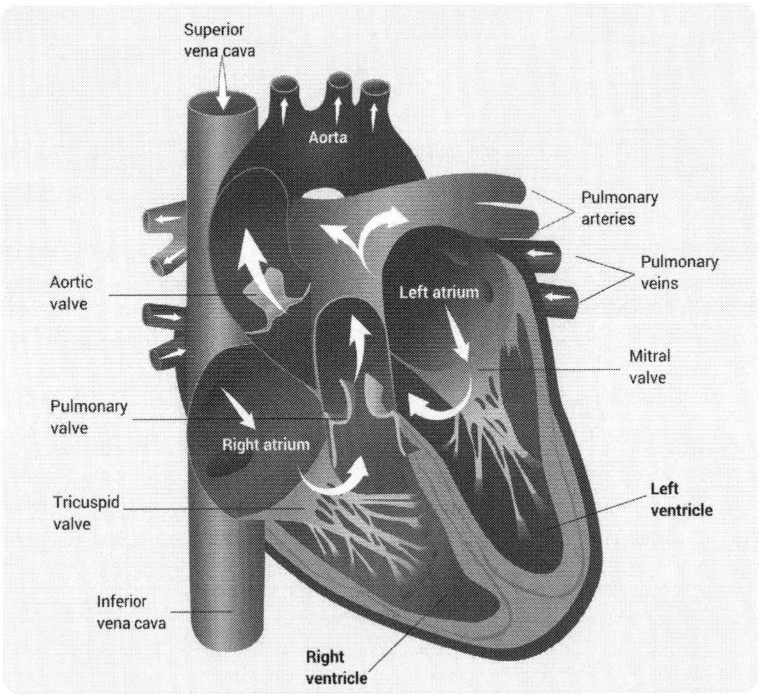

Cardiac Cycle

A cardiac cycle is one complete sequence of cardiac activity. The cardiac cycle represents the relaxation and contraction of the heart and can be divided into two phases: diastole and systole.

Diastole is the phase during which the heart relaxes and fills with blood. It gives rise to the diastolic blood pressure (DBP), which is the bottom number of a blood pressure reading. Systole is the phase during which the heart contracts and discharges blood. It gives rise to the systolic blood pressure (SBP), which is the top number of a blood pressure reading. The heart's electrical conduction system coordinates the cardiac cycle.

Types of Circulation

Five major blood vessels manage blood flow to and from the heart: the superior and inferior venae cavae, the aorta, the pulmonary artery, and the pulmonary vein.

The superior vena cava is a large vein that drains blood from the head and the upper body. The inferior vena cava is a large vein that drains blood from the lower body. The aorta is the largest artery in the human body and carries blood from the heart to body tissues. The pulmonary arteries carry blood from the heart to the lungs. The pulmonary veins transport blood from the lungs to the heart.

In the human body, there are two types of circulation: pulmonary circulation and systemic circulation. Pulmonary circulation supplies blood to the lungs. Deoxygenated blood enters the right atrium of the heart and is routed through the tricuspid valve into the right ventricle. Deoxygenated blood then travels from the right ventricle of the heart through the pulmonary valve and into the pulmonary arteries. The pulmonary arteries carry the deoxygenated blood to the lungs. In the lungs, oxygen is absorbed, and carbon dioxide is released. The pulmonary veins carry oxygenated blood to the left atrium of the heart.

Systemic circulation supplies blood to all other parts of the body, except the lungs. Oxygenated blood flows from the left atrium of the heart through the mitral, or bicuspid, valve into the left ventricle of the heart. Oxygenated blood is then routed from the left ventricle of the heart through the aortic valve and into the aorta. The aorta delivers blood to the systemic arteries, which supply the body tissues. In the tissues, oxygen and nutrients are exchanged for carbon dioxide and other wastes. The deoxygenated blood along with carbon dioxide and wastes enter the systemic veins, where they are returned to the right atrium of the heart via the superior and inferior vena cava.

Gastrointestinal System

The human body relies completely on the digestive system to meet its nutritional needs. After foods and beverages are ingested, the digestive system breaks them down into the component nutrients and absorbs them so that the circulatory system can transport them to other cells to use for growth, energy, and cell repair. These nutrients can be classified as proteins, lipids, carbohydrates, vitamins, and minerals.

The digestive system is thought of chiefly in two parts: the digestive tract (also called the alimentary tract or gastrointestinal tract) and the accessory digestive organs. The digestive tract is the pathway in which food is ingested, digested, absorbed, and excreted. It is composed of the mouth, pharynx, esophagus, stomach, small and large intestines, rectum, and anus. **Peristalsis,** or wave-like contractions of smooth muscle, moves food and wastes through the digestive tract. The accessory digestive organs are the salivary glands, liver, gallbladder, and pancreas.

Mouth and Stomach

The mouth is the entrance to the digestive system. Here, the mechanical and chemical digestion of the food begins. The food is chewed mechanically by the teeth and shaped into a **bolus** by the tongue so that it can be more easily swallowed by the esophagus. The food also becomes more watery and pliable with the addition of saliva secreted from the salivary glands, the largest of which are the parotid glands. The glands also secrete **amylase** in the saliva. Amylase is an enzyme that begins chemical digestion and the breakdown of the carbohydrates in the food into simpler sugars.

The food then moves through the pharynx and down the muscular esophagus to the stomach.

The stomach is a large, muscular sac-like organ at the distal end of the esophagus. Here, the bolus of food is subjected to more mechanical and chemical digestion. As it passes through the stomach, it is physically squeezed and crushed while additional secretions turn it into a watery nutrient-filled liquid that exits into the small intestine as **chyme**.

The stomach secretes a great many substances into the **lumen** of the digestive tract. Some cells produce gastrin, a hormone that prompts other cells in the stomach to secrete a gastric acid composed mostly of hydrochloric acid (HCl). The HCl is at such a high concentration and low pH that it denatures most proteins and degrades a lot of organic matter. The stomach also secretes mucous to form a protective film that keeps the corrosive acid from dissolving the stomach's own cells. Gaps in this mucous layer can lead to peptic ulcers. Finally, the stomach also uses digestive enzymes like proteases and lipases to break down proteins and fats, respectively; although there are some gastric lipases here, the stomach is mostly involved in breaking down proteins.

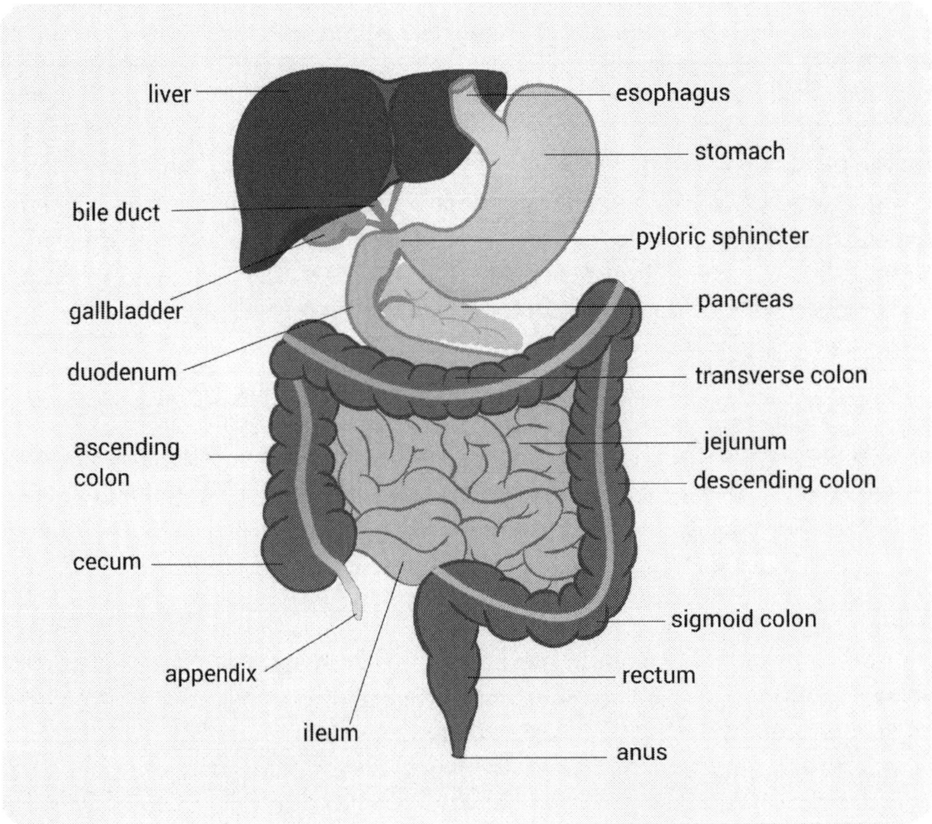

Small Intestine

The chyme from the stomach enters the first part of the small intestine, the **duodenum**, through the pyloric sphincter, and its extreme acidity is partly neutralized by sodium bicarbonate secreted along with mucous. The presence of chyme in the duodenum triggers the secretion of the hormones secretin and cholecystokinin (CCK). Secretin acts on the pancreas to dump more sodium bicarbonate into the small intestine so that the pH is kept at a reasonable level, while CCK acts on the gallbladder to release the bile that it has been storing. Bile is a substance produced by the liver and stored in the gallbladder that helps to **emulsify** or dissolve fats and lipids.

Because of the bile (which aids in lipid absorption) and the secreted lipases (which break down fats), the duodenum is the chief site of fat digestion in the body. The duodenum also represents the last major site of chemical digestion in the digestive tract, as the other two sections of the small intestine (the **jejunum** and **ileum**) are, instead, heavily involved in absorption of nutrients.

The small intestine reaches 40 feet in length, and its cells are arranged in small finger-like projections called **villi**. This is a prime example of where structure aids in function. The small intestine plays a key role in the absorption of nearly all nutrients from the ingested and digested food, effectively transferring them from the lumen of the GI tract to the bloodstream, where they travel to the cells that need them; therefore, the villi increase surface area to allow for more efficient and effective absorption. The absorbed nutrients include simple sugars like glucose from carbohydrates, amino acids from proteins, free fatty acids, electrolytes like sodium and potassium, minerals like iron and zinc, and vitamins like D and B12. Vitamin B12's absorption, though it takes place in the intestines, is actually aided by intrinsic factor, which is released into the chyme back in the stomach prior to reaching the small intestine.

Large Intestine

The leftover parts of food that remain unabsorbed or undigested in the lumen of the small intestine next travel through the large intestine, which is also called the large bowel or colon. The large intestine is mainly responsible for water absorption. At this stage, because the chyme essentially no longer has anything useful that can be absorbed by the body, it is referred to as **waste**, and it is stored in the large intestine until it can be excreted from the body. Removing the liquid from the waste transforms it from liquid to solid stool, or feces.

This waste first passes from the small intestine to the **cecum**, a pouch that forms the first part of the large intestine. In herbivores, it provides a place for bacteria to digest cellulose, but in humans, most of it is vestigial and is known as the **appendix.** From the cecum, waste next travels up the ascending colon, across the transverse colon, down the descending colon, and through the sigmoid colon to the rectum. The rectum is responsible for the final storage of waste before it is expelled through the anus. The anal canal is a small portion of the rectum leading to the anus and the outside of the body.

Pancreas

The pancreas has endocrine and exocrine functions. The endocrine function involves releasing the hormones insulin, which decreases blood sugar (glucose) levels, and glucagon, which increases blood sugar (glucose) levels, directly into the bloodstream. Both hormones are produced in the islets of Langerhans: insulin in the beta cells and glucagon in the alpha cells.

The major part of the gland has exocrine function, which consists of acinar cells secreting inactive digestive enzymes (**zymogens**) into the main pancreatic duct. The main pancreatic duct joins the common bile duct, which empties into the small intestine (specifically the duodenum). The digestive enzymes are then activated and take part in the digestion of carbohydrates, proteins, and fats within chyme.

Neuromuscular System

Nervous System

The human nervous system coordinates the body's response to stimuli from inside and outside the body. There are two major types of nervous system cells: neurons and neuroglia. Neurons are the workhorses of the nervous system and form a complex communication network that transmits electrical impulses termed action potentials, while neuroglia connect and support them.

Although some neurons monitor the senses, some control muscles, and some connect the brain to others, all neurons have four common characteristics:

- Dendrites: These receive electrical signals from other neurons across small gaps called **synapses.**
- Nerve cell body: This is the hub of processing and protein manufacture for the neuron. It may also be called the **soma.**
- Axon: This transmits the signal from the cell body to other neurons.
- Terminals: These bridge the neuron to dendrites of other neurons and deliver the signal via chemical messengers called neurotransmitters.

Here is an illustration of this:

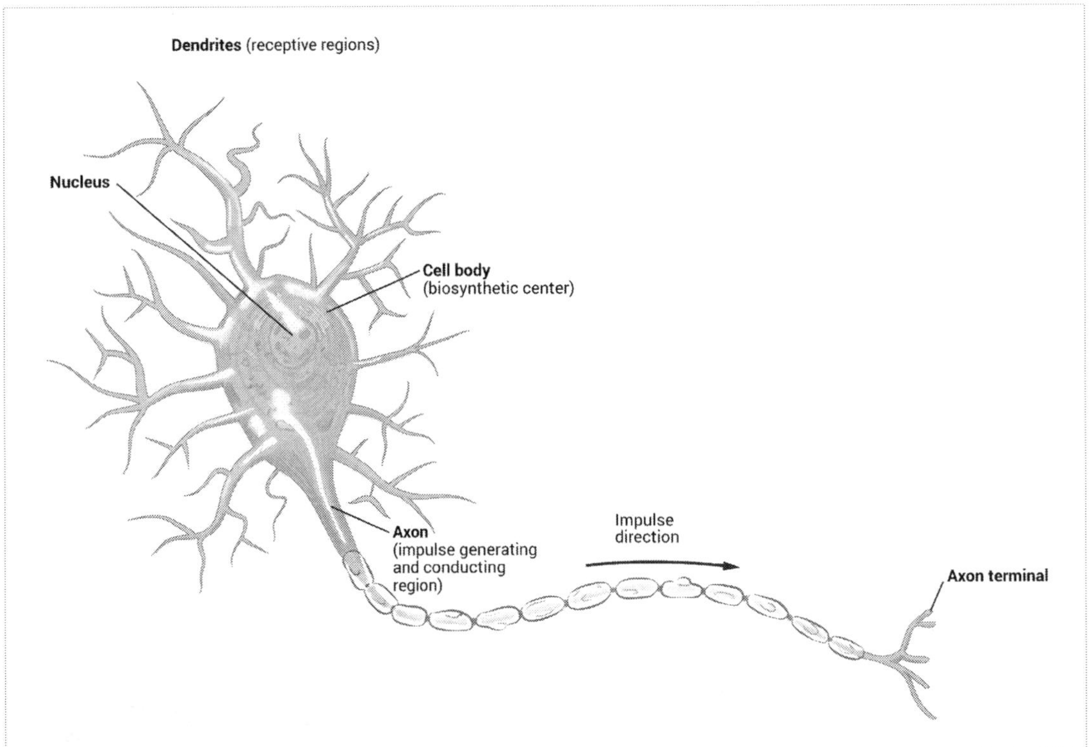

There are two major divisions of the nervous system: central and peripheral:

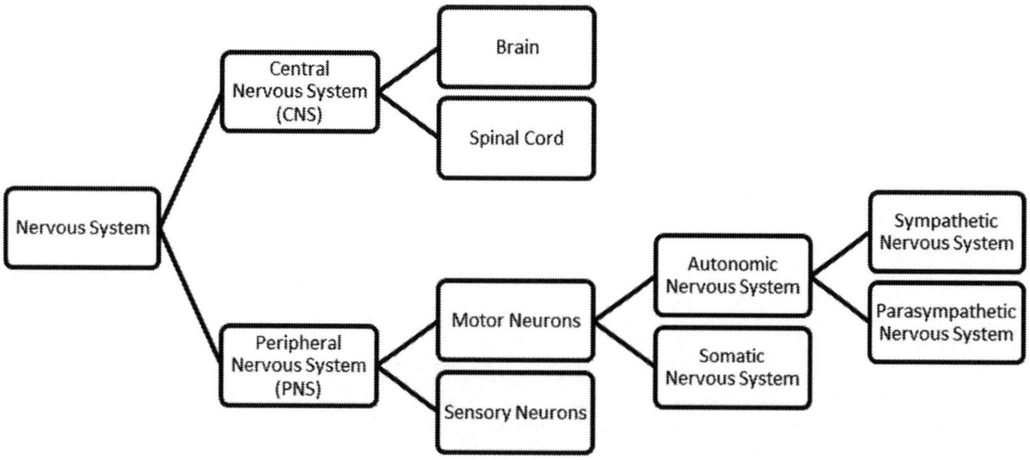

Central Nervous System

The central nervous system (CNS) consists of the brain and spinal cord. Three layers of membranes, called the **meninges**, cover and separate the CNS from the rest of the body.

The major divisions of the brain are the forebrain, the midbrain, and the hindbrain.

The forebrain consists of the cerebrum, the thalamus and hypothalamus, and the rest of the limbic system. The **cerebrum** is the largest part of the brain, and its most well-documented part is the outer cerebral cortex. The cerebrum is divided into right and left hemispheres, and each cerebral cortex hemisphere has four discrete areas, or lobes: frontal, temporal, parietal, and occipital.

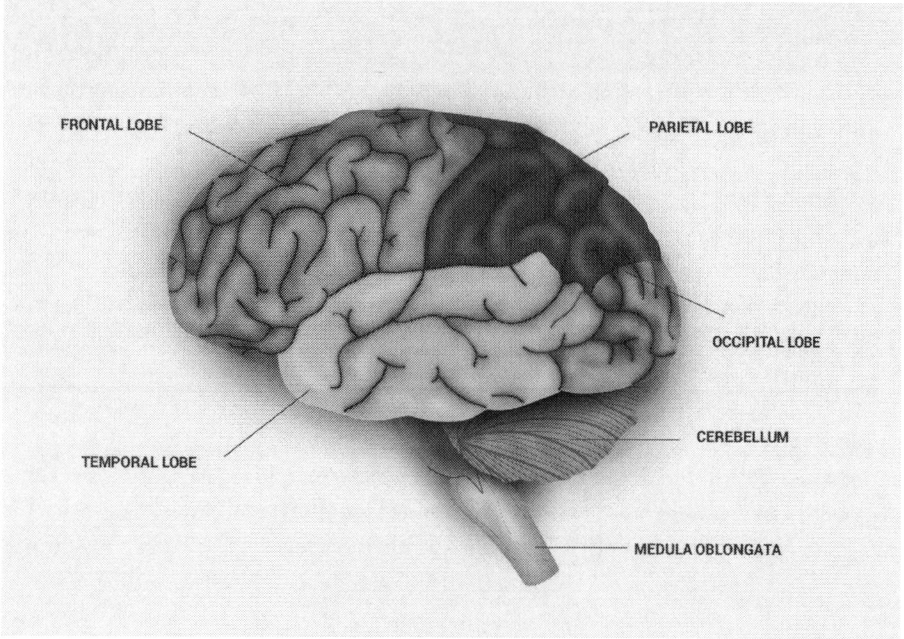

The frontal lobe governs duties such as voluntary movement, judgment, problem solving, and planning, while the other lobes are more sensory in function. The temporal lobe integrates hearing and language comprehension. The parietal lobe processes sensory input from the skin, and integrates taste, temperature, and touch information. The occipital lobe functions to process visual input from the eyes. For completeness, the other two senses, smell and taste, are processed via the olfactory bulbs. The thalamus helps organize and coordinate all of this sensory input in a meaningful way for the brain to interpret.

The hypothalamus controls the endocrine system and all of the hormones that govern long-term effects on the body. Each hemisphere of the limbic system includes a hippocampus (which plays a vital role in memory), an amygdala (which is involved with emotional responses like fear and anger), and other small bodies and nuclei associated with memory and pleasure.

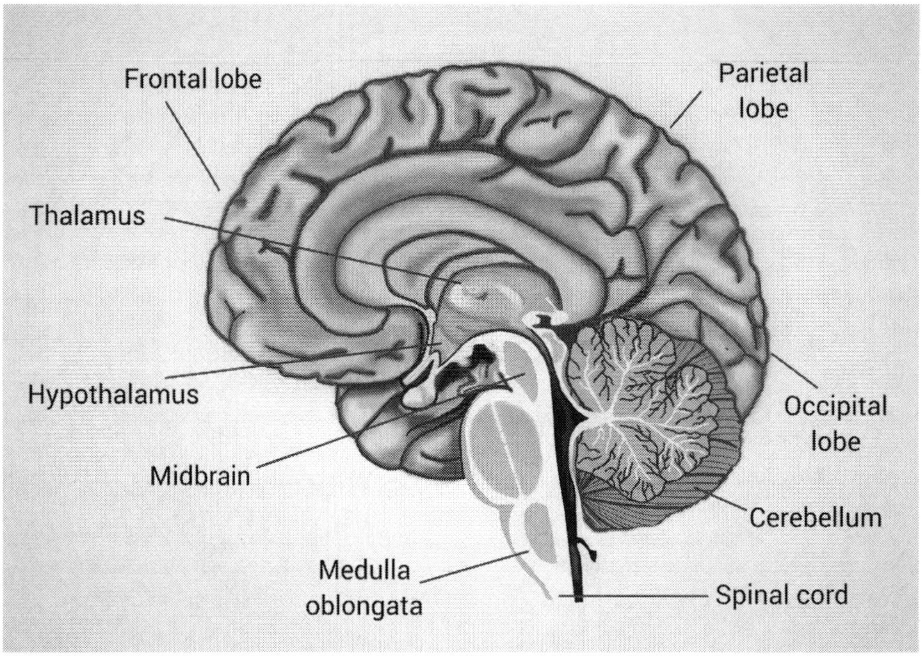

The midbrain is in charge of alertness, sleep/wake cycles, and temperature regulation, and it includes the **substantia nigra**, which produces melatonin to regulate the body's circadian rhythm (sleep/wake cycle). The notable components of the hindbrain include the medulla oblongata and cerebellum. The medulla oblongata is located just above the spinal cord and is responsible for crucial involuntary functions such as breathing, heart rate, swallowing, and the regulation of blood pressure. Together with other parts of the hindbrain, the midbrain and medulla oblongata form the brain stem. The brain stem connects the spinal cord to the rest of the brain. To the rear of the brain stem sits the **cerebellum**, which plays key roles in posture, balance, and muscular coordination. The spinal cord itself carries sensory information to the brain and motor information to the body, encapsulated by its protective bony spinal column.

Peripheral Nervous System

The peripheral nervous system (PNS) includes all nervous tissue besides the brain and spinal cord. The PNS consists of the sets of cranial and spinal nerves and relays information between the CNS and the rest of the body. The PNS has two divisions: the autonomic nervous system and the somatic nervous system.

Autonomic Nervous System

The autonomic nervous system (ANS) governs involuntary, or reflexive, body functions. Ultimately, the autonomic nervous system controls functions such as breathing, heart rate, digestion, body temperature, and blood pressure.

The ANS is split between parasympathetic nerves and sympathetic nerves. These two nerve types are antagonistic, so they have opposite effects on the body. Parasympathetic nerves typically are involved when resting or during safe conditions and decrease heart rate, decrease inhalation speed, prepare digestion, and allow urination and excretion. Sympathetic nerves, on the other hand, become active when a person is under stress or excited, and they increase heart rate, increase respiration, and inhibit digestion, urination, and excretion.

Somatic Nervous System and the Reflex Arc

The somatic nervous system (SNS) governs the conscious, or voluntary, control of skeletal muscles and their corresponding body movements. The SNS contains afferent and efferent neurons. Afferent neurons carry sensory messages from the skeletal muscles, skin, or sensory organs to the CNS. Efferent neurons relay motor messages from the CNS to skeletal muscles, skin, or sensory organs.

The SNS also has a role in involuntary movements called reflexes. A **reflex** is defined as an involuntary response to a stimulus. They are transmitted via what is termed a **reflex arc,** where a stimulus is sensed by an affector and its afferent neuron, interpreted and rerouted by an interneuron, and delivered to effector muscles by an efferent neuron where they respond to the initial stimulus. A reflex is able to bypass the brain by being rerouted through the spinal cord; the interneuron decides the proper course of action rather than the brain. The reflex arc results in an instantaneous, involuntary response. For example, a physician tapping on the knee produces an involuntary knee jerk referred to as the patellar tendon reflex.

Muscular System

The muscular system is responsible for involuntary and voluntary movement of the body. There are three types of muscle: skeletal, cardiac, and smooth.

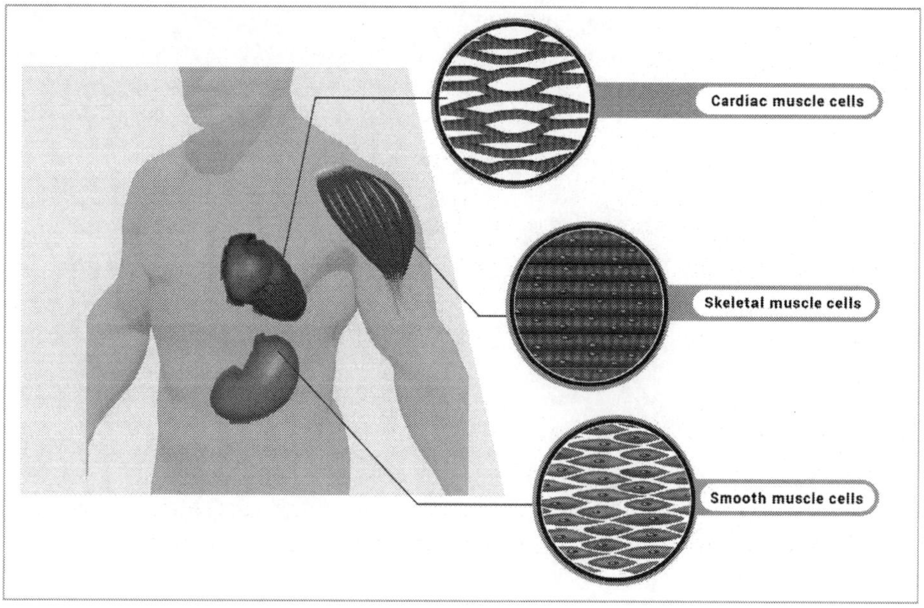

Skeletal Muscles

Skeletal muscles, or voluntary muscles, are attached to bones by tendons and are responsible for voluntary movement. The connecting tendons are made up of dense bands of connective tissue and have collagen fibers that firmly attach the muscle to the bone. Their fibers are actually woven into the coverings of the bone and muscle so that they can withstand pressure and tension. They usually work in opposing pairs like the quadriceps on the front of the thigh and the hamstrings on the back of the thigh, for example.

Skeletal muscles are made of bundles of long fibers that are composed of cells with many nuclei due to their length. These fibers contain myofibrils, and myofibrils are made of alternating filaments. The thicker myosin filaments are in between the smaller actin filaments in a unit called a **sarcomere**, and the overlapping regions give the muscle their characteristic striated, or striped, appearance. Actin filaments are attached to exterior Z lines, myosin filaments are attached to a central M line, and when a muscle is at rest, there is a gap between the Z line and the myosin filaments. Only when the muscle contracts and the actin filaments slide over the myosin filaments does the myosin reach the Z line, as illustrated in the picture below. This sliding-filament model of muscle contraction is dependent on myosin molecules forming and breaking cross-bridges with actin in order to pull the actin filaments closer to the M line.

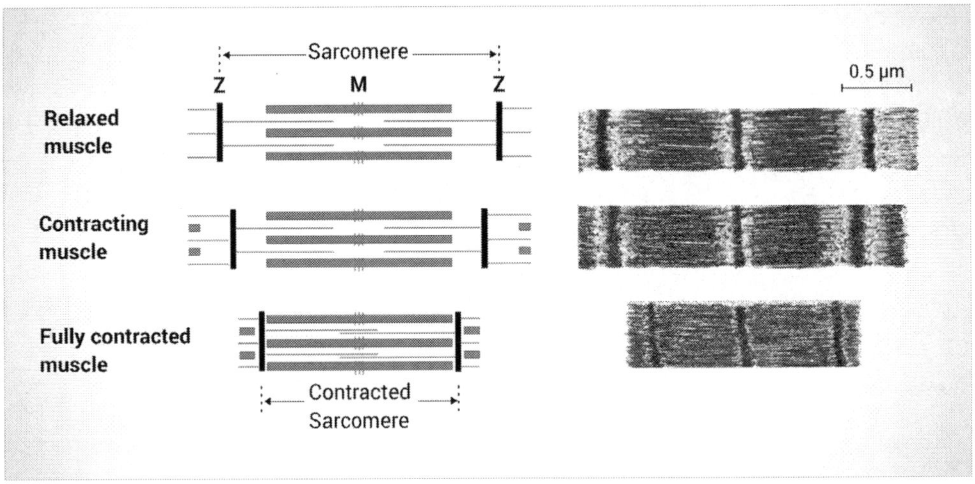

Skeletal muscles are controlled by the nervous system. Motor neurons connect to muscle fibers via neuromuscular junctions. Motor neurons must release the neurotransmitter acetylcholine, which releases calcium ions to stimulate myosin cross-bridging and contraction. As the acetylcholine stops being released, the contraction ends.

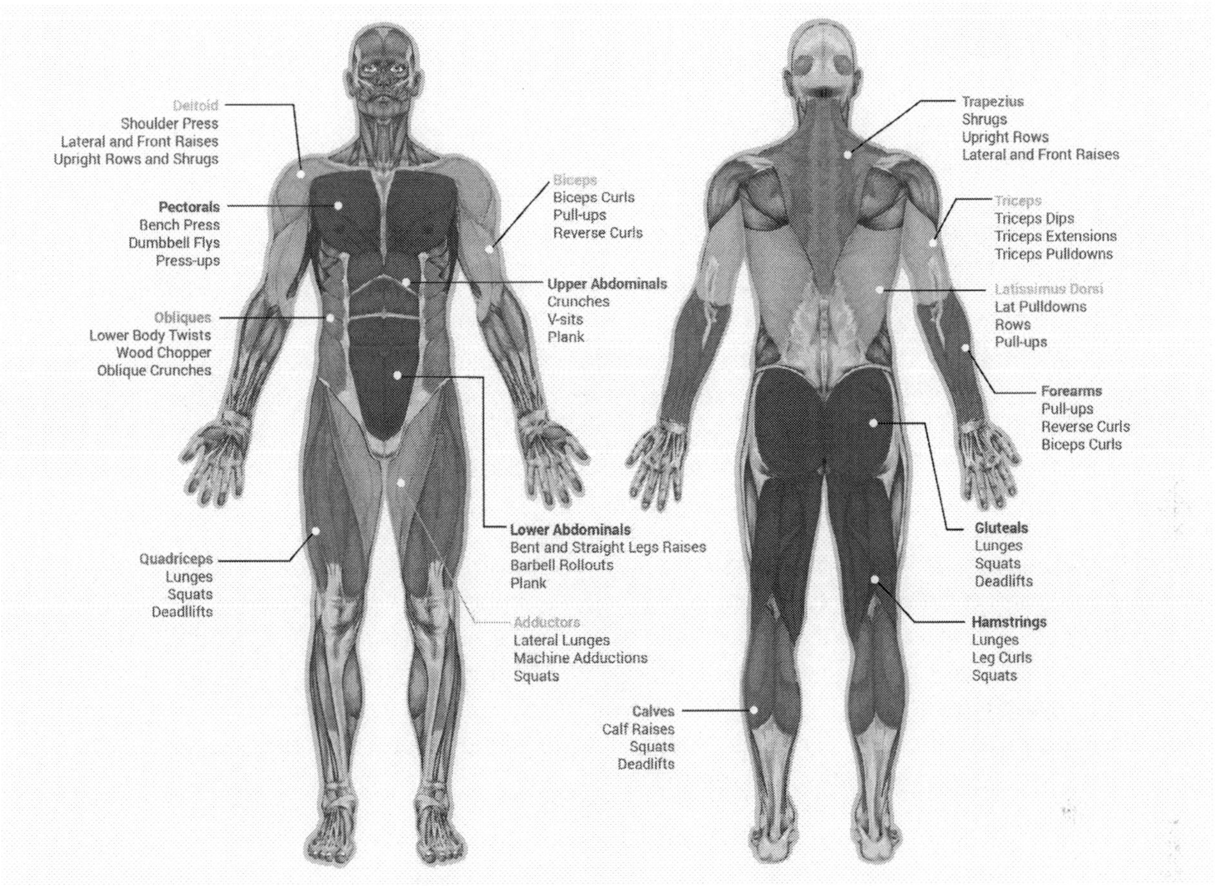

<u>Smooth Muscles</u>

Smooth muscles are responsible for involuntary movement like food moving through the digestive tract and blood moving through vessels. They have only one nucleus and do not have striations because actin and myosin filaments do not have an organized arrangement like they do in skeletal muscles. Unlike skeletal muscle, these muscles don't require neuromuscular junctions. Instead, they operate via gap junctions, which send impulses directly from cell to cell.

<u>Cardiac Muscles</u>

Cardiac muscle cells are found only in the heart, where they control the heart's rhythm and blood pressure. Like skeletal muscle, cardiac muscle has striations, but cardiac muscle cells are smaller than skeletal muscle cells, so they typically have only one nucleus. Like smooth muscle, cardiac muscles do not require the release of neurotransmitters by motor neurons to function, and they instead operate via gap junctions.

Reproductive System

The reproductive system is responsible for producing, storing, nourishing, and transporting functional reproductive cells, or **gametes**, in the human body. It includes the reproductive organs, also known as

gonads, the reproductive tract, the accessory glands and organs that secrete fluids into the reproductive tract, and the perineal structures, which are the external genitalia.

Reproduction involves the passing of genes from one generation to the next, and that is accomplished through haploid gametes. Gametes have gone through meiosis and have 23 chromosomes, half the normal number. The male gamete is sperm, and the female gamete is an egg or ovum. When a sperm fertilizes an egg, they create a zygote, which is the first cell of a new organism. The zygote has a full set of 46 chromosomes because it received 23 from each parent. Because of crossing over, which occurs during meiosis, sperm and egg chromosome sets are all different, which results in the variety and genetic diversity seen in humans.

Male Reproductive System

The entire male reproductive system is designed to generate sperm and produce semen, a lubricant that facilitates the fertilization of eggs, the female gametes. The testes are the endocrine glands that secrete testosterone, a hormone that is important for secondary sex characteristics and sperm development, or **spermatogenesis**. Testosterone is in the androgen steroid-hormone family. The testes also produce and store 500 million spermatocytes (sperm), which are the male gametes, each day. Testes are housed in the scrotum, which is a sac that hangs outside the body so that spermatogenesis occurs at cooler and optimal conditions.

The seminiferous tubules within the testes produce sperm, which then travel to epididymis where they are stored as they mature. Then, the sperm move to the ejaculatory duct via the vas deferens. The ejaculatory duct contains more than just sperm. The seminal vesicles secrete an alkaline substance that will help sperm survive in the acidic vagina. The prostate gland secretes enzymes bathed in a milky white fluid that is important for thinning semen after ejaculation to increase its likelihood of reaching the egg. The bulbourethral, or Cowper's, gland secretes an alkaline fluid that lubricates the urethra prior to ejaculation to neutralize any acidic urine residue.

The sperm, along with all the exocrine secretions, are collectively called **semen**. Their destination is the vagina and they can only get there if the penis is erect from arousal and increased circulation. During sexual intercourse, ejaculation will forcefully expel the contents of the semen and effectively deliver the sperm to the egg. The muscular prostate gland is important for ejaculation. Each ejaculation releases 2 to 6 million sperm. Sperm has a whip-like flagellum tail that facilitates movement and helps propel it toward the egg.

Female Reproductive System

The vagina is the passageway that sperm must travel through to reach an egg, the female gamete. Surrounding the vagina are the labia minor and labia major, both of which are folds that protect the urethra and the vaginal opening. The clitoris is rich in nerve-endings, making it sensitive and highly stimulated during sexual intercourse. It is above the vagina and urethra. The Bartholin's glands secretes a fluid during arousal that is important for lubrication. These pea-sized glands are exocrine glands.

The female gonads are the ovaries. Ovaries generally produce one immature gamete, an egg or oocyte, per month. They are also responsible for secreting the hormones estrogen and progesterone. Fertilization cannot happen unless the ejaculated sperm finds the egg, which is only available for a brief period of the menstrual cycle. Eggs, or ova, develop in the ovaries in clusters surrounded by follicles, and after puberty, one is delivered to the uterus per month via the fallopian tubes. The 28-day average journey of the egg to the uterus is called the **menstrual cycle** and it is highly regulated by the endocrine system. The regulatory hormones Gonadotropin releasing hormone (GnRH), luteinizing hormone (LH), and follicle-stimulating hormone (FSH) orchestrate the menstrual cycle. Ovarian hormones estrogen and progesterone are also important in timing as well as for vascularization of the uterus in preparation for pregnancy. Fertilization usually happens around **ovulation**, which is when the egg is inside the fallopian tube. The resulting zygote travels down the tube and implants into the uterine wall. The uterus protects and nourishes the developing embryo for nine months until it is ready for the outside environment.

If the egg released is unfertilized, the uterine lining will slough off during menstruation. Should a fertilized egg, called a **zygote**, reach the uterus, it will embed itself into the uterine wall due to uterine vascularization, which will deliver blood, nutrients, and antibodies to the developing embryo. The uterus is where the embryo will develop for the next nine months. Mammary glands are important female reproductive structures because they produce milk they provide for their young during lactation. Milk contains nutrients and antibodies that benefit the baby.

Integumentary System

The integumentary system includes skin, hair, nails, oil glands, and sweat glands. The largest organ of the integumentary system (and of the body), the skin, acts as a barrier and protects the body from mechanical impact, variations in temperature, microorganisms, chemicals, and UV radiation from the sun. It regulates body temperature, peripheral circulation, and excretes waste through sweat. It also contains a large network of nerve cells that relay changes in the external environment to the brain.

Layers of Skin

Skin consists of two layers: the superficial epidermis and the deeper dermis. The subcutaneous hypodermis is below the dermis and contains a layer of fat and connective tissue that are both important for insulation.

The whole epidermis is composed of epithelial cells that lack blood vessels. The outer epidermis is composed of dead cells which surround the living cells underneath. The most inner epidermal tissue is a single layer of cells called the **stratum basale,** which is composed of rapidly-dividing cells that push old cells to the skin's surface. When being pushed out, the cells' organelles disappear and they start producing a protein called keratin that eventually forms a tough waterproof layer. This outer layer sloughs off every four to five weeks. The melanocytes in the stratum basale produce the pigment melanin that absorbs UV rays and protects the skin. Skin also produces vitamin D if exposed to sunlight.

The dermis underneath the epidermis contains supporting collagen fibers peppered with nerves, blood vessels, hair follicles, sweat glands, oil glands, and smooth muscles.

Skin's Involvement in Temperature Homeostasis

The skin has a thermoregulatory role in the human body that is controlled by a negative feedback loop. The control center of temperature regulation is the hypothalamus in the brain. When the hypothalamus is alerted by receptors from the dermis, it secretes hormones that activate effectors to keep internal temperature at a set point of 98.6°F (37°C). If the environment is too cold, the hypothalamus will initiate a pathway that induces muscle shivering to release heat energy as well as constrict blood vessels to limit heat loss. In hot conditions, the hypothalamus will initiate a pathway that vasodilates blood vessels to increase heat loss and stimulate sweating for evaporative cooling. Evaporative cooling occurs when the hottest water particles evaporate and leave behind the coolest ones. This cools down the body.

Sebaceous Glands vs. Sweat Glands

The skin also contains oil glands, or sebaceous glands, and sweat glands that are exocrine because their substances are secreted through ducts. Endocrine glands secrete substances into the bloodstream instead. Oil glands are attached to hair follicles. They secrete **sebum**, an oily substance that moisturizes the skin, protecting it from water loss. Sebum also keeps the skin elastic. Also, sebum's slight acidity provides a chemical defense against bacterial and fungal infections.

Sweat glands not attached to hair follicles are called eccrine glands. They are all over the body and these are the ones responsible for thermoregulation. They also remove bodily waste by secreting water and electrolytes. Sweat glands attached to hair follicles are apocrine glands and there are not nearly as many of these sweat glands. Apocrine glands are only active post-puberty. They secrete a thicker, viscous substance that is attractive to bacteria, leading to the unpleasant smell in armpits, feet, and the groin. They are stimulated during stress and arousal.

Endocrine System

The endocrine system is made up of the ductless tissues and glands that secrete hormones directly into the bloodstream. It is similar to the nervous system in that it controls various functions of the body, but it does so via secretion of hormones in the bloodstream as opposed to nerve impulses. The endocrine system is also different because its effects last longer than that of the nervous system. Nerve impulses are immediate, while hormone responses can last for minutes or even days.

The endocrine system works closely with the nervous system to regulate the physiological activities of the other systems of the body in order to maintain homeostasis. Hormone secretions are controlled by tight feedback loops that are generally regulated by the hypothalamus, the bridge between the nervous and endocrine systems. The hypothalamus receives sensory input via the nervous system and responds by stimulating or inhibiting the pituitary gland, which, in turn, stimulates or inhibits several other glands. The tight control is due to hormone secretions.

Hormones are chemicals that bind to specific target cells. Each hormone will only bind to a target cell that has a specific receptor with the correct shape. For example, testosterone will not attach to skin cells because skin cells do not have any receptors that recognize testosterone.

There are two types of hormones: steroid and protein. Steroid hormones are lipid, nonpolar substances, and most are able to diffuse across cell membranes. Once they do, they bind to a receptor that initiates a signal transduction cascade that affects gene expression. Non-steroid hormones bind to receptors on cell membranes that also initiate a signal transduction cascade that affects enzyme activity and chemical reactions.

Major Endocrine Glands

Hypothalamus: This gland is a part of the brain. It connects the nervous system to the endocrine system because it receives sensory information through nerves and it sends instructions via hormones delivered to the pituitary.

Pituitary Gland: This gland is pea-sized and is found at the bottom of the hypothalamus. It has two lobes called the anterior and posterior lobes. It plays an important role in regulating other endocrine glands. For example, it secretes growth hormone which regulates growth. Other hormones that are released by this gland control the reproductive system, childbirth, nursing, blood osmolarity, and metabolism.

The hormones and glands respond to each other via feedback loops, and a typical feedback loop is illustrated in the picture below. The hypothalamus and pituitary gland are master controllers of most of the other glands.

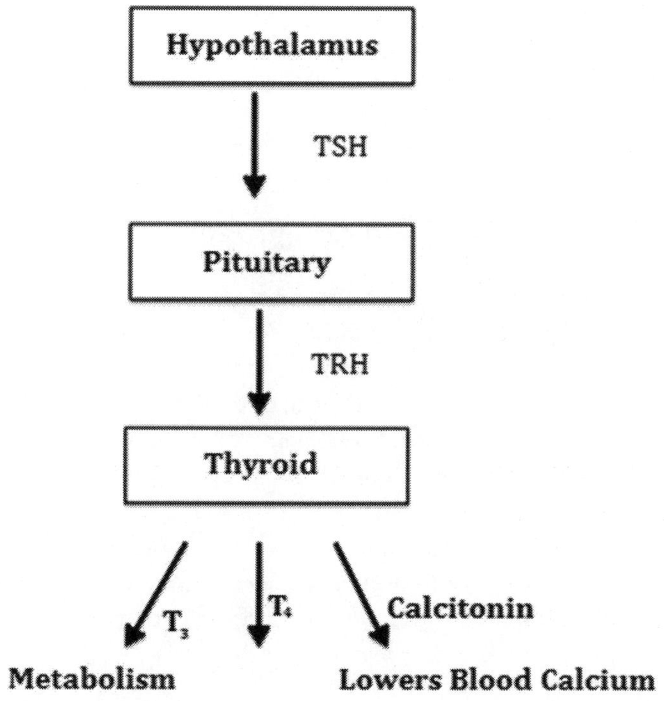

Thymus Gland: This gland is located in the chest cavity, embedded in connective tissue. It produces several hormones that are important for development and maintenance of T lymphocytes, which are important cells for immunity.

Adrenal Glands: One adrenal gland is attached to the top of each kidney. The adrenal glands produce epinephrine and norepinephrine, which cause the "fight or flight" response in the face of danger or stress. These hormones raise heart rate, blood pressure, dilate bronchioles, and deliver blood to the muscles. All of these actions increase circulation and release glucose so that the body has an energy burst.

Pineal Gland: The pineal gland secretes melatonin, which is a hormone that regulates the body's circadian rhythm (the natural wake-sleep cycle).

Testes and Ovaries: They secrete testosterone and both estrogen and progesterone, respectively. They are responsible for secondary sex characteristics, gamete development, and female hormones are important for embryonic development.

Thyroid Gland: This gland releases hormones like thyroxine and calcitonin. Thyroxine stimulates metabolism, and calcitonin monitors the amount of circulating calcium. Calcitonin signals the body to regulate calcium from bone reserves and kidney reabsorption of calcium.

Parathyroid Glands: These are four pea-sized glands located on the posterior surface of the thyroid. The main hormone that is secreted is called parathyroid hormone (PTH), which influences calcium levels much like calcitonin does, except PTH is antagonistic: PTH increases extracellular levels of calcium while calcitonin decreases it.

Pancreas: The pancreas is an organ that has both endocrine and exocrine functions. It functions outside of a typical feedback loop in that blood sugar concentration seems to signal the pancreas itself. The endocrine functions are controlled by the pancreatic islets of Langerhans, which are groups of beta cells scattered throughout the gland that secrete insulin to lower blood sugar levels in the body. Neighboring alpha cells secrete glucagon to raise blood sugar. These complementary hormones keep blood sugar in check.

Genitourinary System

The urinary system is made up of the kidneys, ureters, urinary bladder, and the urethra. It is the system responsible for removing waste products and balancing water and electrolyte concentrations in the blood. The urinary system has many important functions related to waste excretion. It regulates the concentrations of sodium, potassium, chloride, calcium, and other ions in the filtrate by controlling the amount of each that is reabsorbed during filtration. The reabsorption or secretion of hydrogen ions and bicarbonate contributes to the maintenance of blood pH. Certain kidney cells can detect any reductions in blood volume and pressure. If that happens, they secrete renin, which will activate a hormone that causes increased reabsorption of sodium ions and water, raising volume and pressure. Under hypoxic conditions, kidney cells will secrete erythropoietin in order to stimulate red blood cell production. Kidney cells also synthesize calcitriol, which is a hormone derivative of vitamin D3 that aids in calcium ion absorption by the intestinal epithelium.

Under normal circumstances, humans have two functioning kidneys in the lower back on either side of the spinal cord. They are the main organs that are responsible for filtering waste products out of the blood and regulating blood water and electrolyte levels. Blood enters the kidney through the renal artery and urea and wastes are removed while water and the acidity/alkalinity of the blood is adjusted. Toxic substances and drugs are also filtered. Blood exits through the renal vein and the urine waste travels through the ureter to the bladder where it is stored until it is eliminated through the urethra.

The kidneys have an outer renal cortex and an inner renal medulla that contain millions of tiny filtering units called **nephrons**. Nephrons have two parts: a glomerulus, which is the filter, and a tubule. The **glomerulus** is a network of capillaries covered by the Bowman's capsule, which is the entrance to the tubule. As blood enters the kidneys via the renal artery, the glomerulus allows for fluid and waste products to pass through it and enter the tubule. Blood cells and large molecules, such as proteins, do not pass through and remain in the blood. The filtrate passes through the tubule, which has several parts. The proximal tubule comes first, and then the descending and ascending limbs of the loop of Henle dip into the medulla, followed by the distal tubule and collecting duct. The journey through the tubule involves a balancing act that regulates blood osmolarity, pH, and electrolytes exchange of materials between the tubule and the blood stream. The final product at the collecting tubule is called **urine**, and it is delivered to the bladder by the ureter. The most central part of the kidney is the renal pelvis, and it acts as a funnel by delivering the urine from the millions of the collecting tubules to the ureters. The filtered blood exits through the renal vein and is returned to circulation.

Here's a look at the genitourinary system:

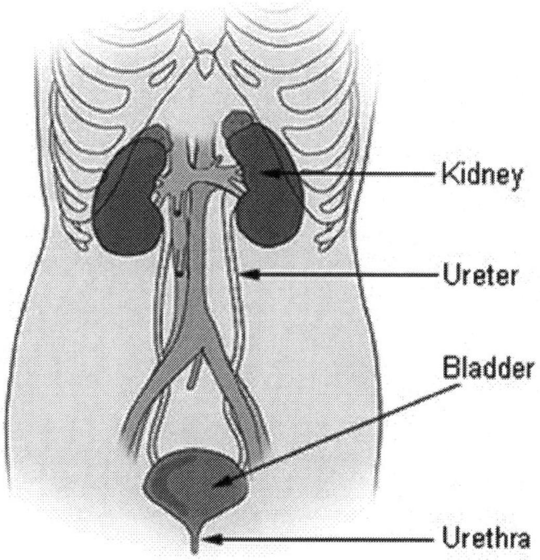

Here's a close up look at the kidney:

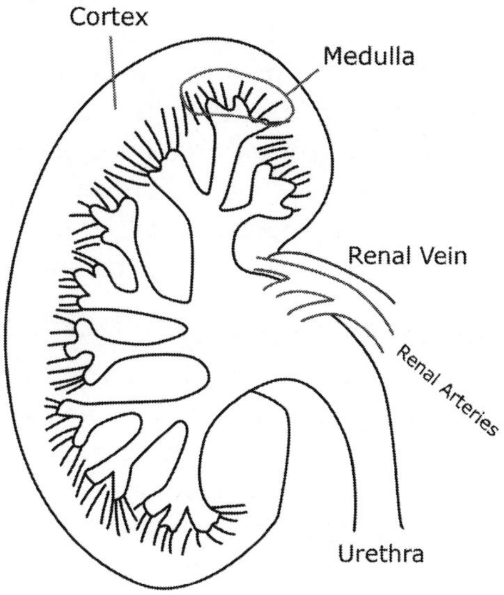

Waste Excretion

Once urine accumulates, it leaves the kidneys. The urine travels through the ureters into the urinary bladder, a muscular organ that is hollow and elastic. As more urine enters the urinary bladder, its walls stretch and become thinner so there is no significant difference in internal pressure. The urinary bladder stores the urine until the body is ready for urination, at which time the muscles contract and force the urine through the urethra and out of the body.

Immune System

The immune system is the body's defense against invading microorganisms (bacteria, viruses, fungi, and parasites) and other harmful, foreign substances. It is capable of limiting or preventing infection.

There are two general types of immunity: innate immunity and acquired immunity. **Innate immunity** uses physical and chemical barriers to block microorganism entry into the body. The biggest barrier is the skin; it forms a physical barrier that blocks microorganisms from entering underlying tissues. Mucous membranes in the digestive, respiratory, and urinary systems secrete mucus to block and remove invading microorganisms. Other natural defenses include saliva, tears, and stomach acids, which are all chemical barriers intended to block infection with microorganisms. Acid is inhospitable to pathogens, as are tears, mucus, and saliva which all contain a natural antibiotic called lysozyme. The respiratory passages contain microscopic cilia which are like bristles that sweep out pathogens. In addition, macrophages and other white blood cells can recognize and eliminate foreign objects through phagocytosis or toxic secretions.

Acquired immunity refers to a specific set of events used by the body to fight a particular infection. Essentially, the body accumulates and stores information about the nature of an invading microorganism. As a result, the body can mount a specific attack that is much more effective than innate immunity. It also provides a way for the body to prevent future infections by the same microorganism.

Acquired immunity is divided into a primary response and a secondary response. The primary immune response occurs the first time a particular microorganism enters the body, where macrophages engulf the microorganism and travel to the lymph nodes. In the lymph nodes, macrophages present the invader to helper T lymphocytes, which then activate humoral and cellular immunity. **Humoral immunity** refers to immunity resulting from antibody production by B lymphocytes. After being activated by helper T lymphocytes, B lymphocytes multiply and divide into plasma cells and memory cells. Plasma cells are B lymphocytes that produce immune proteins called antibodies, or immunoglobulins. Antibodies then bind the microorganism to flag it for destruction by other white blood cells. **Cellular immunity** refers to the immune response coordinated by T lymphocytes. After being activated by helper T lymphocytes, other T lymphocytes attack and kill cells that cause infection or disease.

The secondary immune response takes place during subsequent encounters with a known microorganism. Memory cells respond to the previously encountered microorganism by immediately producing antibodies. Memory cells are B lymphocytes that store information to produce antibodies. The secondary immune response is swift and powerful, because it eliminates the need for the time-consuming macrophage activation of the primary immune response. Suppressor T lymphocytes also take part to inhibit the immune response as an overactive immune response could cause damage to healthy cells.

Inflammation occurs if a pathogen evades the barriers and chemical defenses. It stimulates pain receptors, alerting the individual that something is wrong. It also elevates body temperature to speed up chemical reactions, although if a fever goes unchecked it can be dangerous due to the fact that extreme heat unfolds proteins. Histamine is secreted; it dilates blood vessels and recruits white blood cells that destroy invaders non-specifically. The immune system is tied to the lymphatic system. The thymus, one of the lymphatic system's organs, is the site of maturation of T-cells, a type of white blood cell. The lymphatic system is important in the inflammatory response because lymph vessels deliver leukocytes and collect debris that will be filtered in the lymph nodes and the spleen.

Antigen and Typical Immune Response

Should a pathogen evade barriers and survive through inflammation, an antigen-specific adaptive immune response will begin. Immune cells recognize these foreign particles by their antigens, which are their unique and identifying surface proteins. Drugs, toxins, and transplanted cells can also act as antigens. The body even recognizes its own cells as potential threats in autoimmune diseases.

When a macrophage engulfs a pathogen and presents its antigens, helper T cells recognize the signal and secrete cytokines to signal T lymphocytes and B lymphocytes so that they launch the cell-mediated and humoral response, respectively. The cell-mediated response occurs when the T lymphocytes kill infected cells by secreting cytotoxins. The humoral response occurs when B lymphocytes proliferate into plasma and memory cells. The plasma cells secrete antigen-specific antibodies, which bind to the pathogens so that they cannot bind to host cells. Macrophages and other phagocytic cells called **neutrophils** engulf and degrade the antibody/pathogen complex. The memory cells remain in circulation and initiate a secondary immune response should the pathogen dare enter the host again.

Active and Passive Immunity

Acquired immunity occurs after the first antigen encounter. As mentioned, the first time the body mounts this immune response is called the primary immune response. Because the memory B cells store information about the antigen's structure, any subsequent immune response causes a secondary immune response, which is much faster and substantially more antibodies are produced due to the presence of memory B cells. If the secondary immune response is strong and fast enough, it will fight off the pathogen before an individual becomes symptomatic. This is a natural means of acquiring immunity.

Vaccination is the process of inducing immunity. **Active immunization** refers to immunity gained by exposure to infectious microorganisms or viruses and can *be* natural or artificial. **Natural immunization** refers to an individual being exposed to an infectious organism as a part of daily life. For example, it was once common for parents to expose their children to childhood diseases such as measles or chicken pox. **Artificial immunization** refers to therapeutic exposure to an infectious organism as a way of protecting an individual from disease. Today, the medical community relies on artificial immunization as a way to induce immunity.

Vaccines are used for the development of active immunity. A vaccine contains a killed, weakened, or inactivated microorganism or virus that is administered through injection, by mouth, or by aerosol. Vaccinations are administered to prevent an infectious disease but do not always guarantee immunity. Due to circulating memory B cells after administration, the secondary response will fight off the pathogen should it be encountered again in many cases. Both illnesses and vaccinations cause active immunity.

Passive immunity refers to immunity gained by the introduction of antibodies. This introduction can also be natural or artificial. The process occurs when antibodies from the mother's bloodstream are passed on to the bloodstream of the developing fetus. Breast milk can also transmit antibodies to a baby. Babies are born with passive immunity, which provides protection against general infection for approximately the first six months of its life.

Types of Leukocytes

There are many leukocytes, or white blood cells, involved in both innate and adaptive immunity. All are developed in bone marrow. Many have been mentioned in the text above, but a comprehensive list is included here for reference.

- Monocytes are large phagocytic cells.
    - Macrophages engulf pathogens and present their antigen. Some circulate, but others reside in lymphatic organs like the spleen and lymph nodes.
    - Dendritic cells are also phagocytic and antigen-presenting.
- Granulocytes are cells that contain secretory granules.
    - Neutrophils are the most abundant white blood cell. They are circulating and aggressive phagocytic cells that are part of innate immunity. They also secrete substances that are toxic to pathogens.
    - Basophils and mast cells secrete histamine which stimulates the inflammatory response.
    - Eosinophils are found underneath mucous membranes and defend against multicellular parasites like worms. They have low phagocytic activity and primarily secrete destructive enzymes.
- T lymphocytes mature in the thymus.
    - Helper T cells recognize antigens presented by macrophages and dendritic cells and secrete cytokines that mount the humoral and cell-mediated immune response.
    - Killer T cells are cytotoxic cells involved in the cell-mediated response by recognizing and poisoning infected cells.
    - Suppressor T cells suppress the adaptive immune response when there is no threat to conserve resources and energy.
    - Memory T cells remain in circulation to aid in the secondary immune response.
- B lymphocytes mature in bone marrow.
    - Plasma B cells secrete antigen-specific antibodies when signaled by Helper T cells and are degraded after the immune response.
    - Memory B cells store antigen-specific antibody making instructions and remain circulating after the immune response is over.
- Natural killer cells are part of innate immunity and patrol and identify suspect-material. They respond by secreting cytotoxic substances.

Skeletal System

Axial Skeleton and Appendicular Skeleton

The skeletal system is composed of 206 bones interconnected by tough connective tissue called ligaments. The axial skeleton can be considered the north-south axis of the skeleton. It includes the spinal column, sternum, ribs, and skull. There are 80 bones in the axial skeleton, and 33 of them are vertebrae. The ribs make up 12 of the bones in the axial skeleton.

The remaining 126 bones are in the appendicular skeleton which contains bones of the appendages like the collarbone (clavicle), shoulders (scapula), arms, hands, hips, legs, and feet. The arm bones consist of the upper humerus with the radius and ulna that attach to the hands. The wrists, hands, and fingers are composed of the carpals, metacarpals, and phalanges, respectively. The femur attaches to the hips. The patella or kneecap connects the femur to the fibula and tibia. The ankles, feet, and toes are composed of the tarsals, metatarsals, and phalanges, respectively.

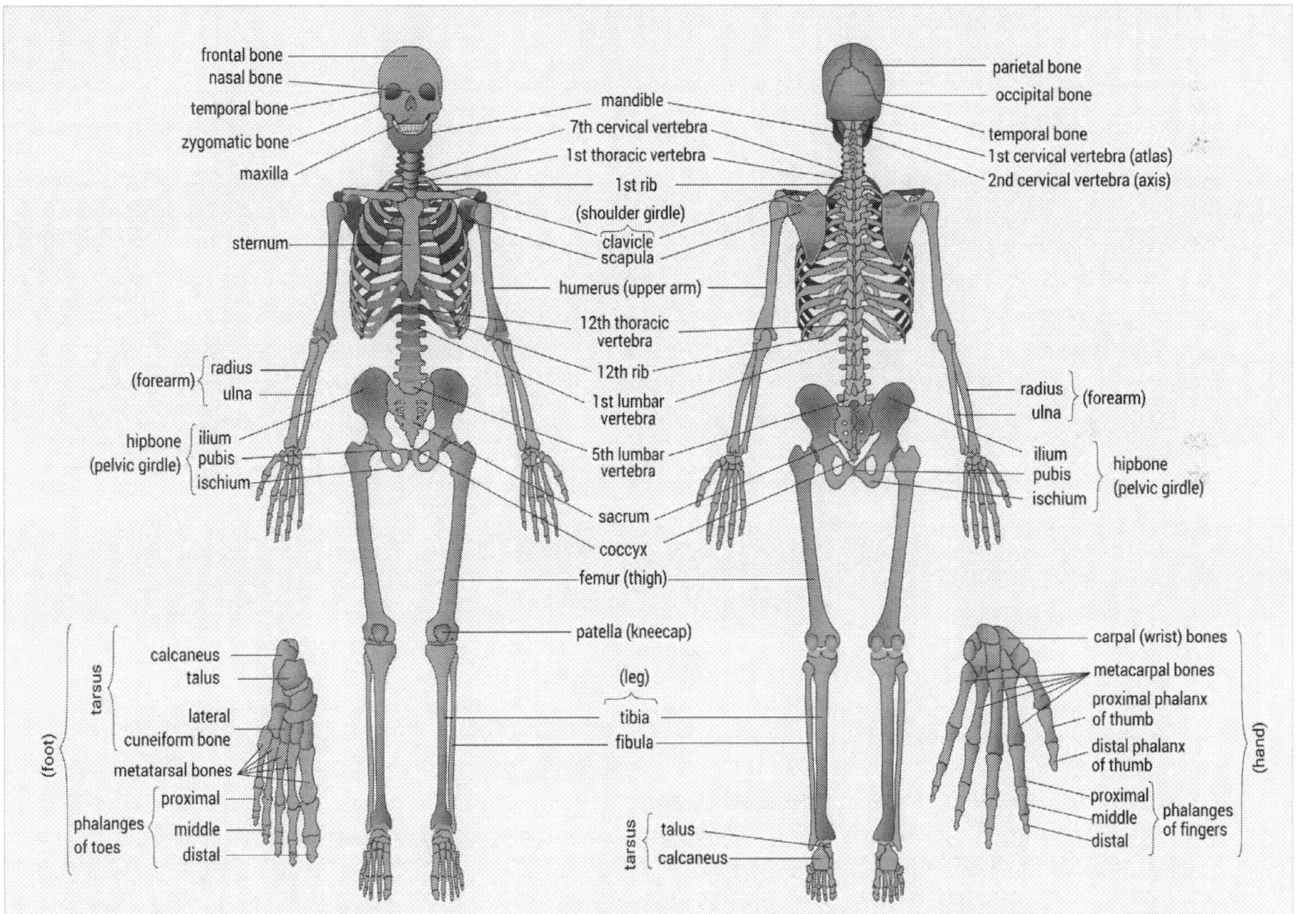

Functions of the Skeletal System

One of the skeletal system's most important functions is to protect vital internal organs. The skull protects the brain, the ribs protect the lungs and heart, the vertebrae protect the spinal column, and the pelvis protects the uterus and other reproductive organs. Red bone marrow produces red and white blood cells as well as platelets in a process known as hemopoiesis. The bones themselves store the essential minerals calcium and phosphorous. The organization of the skeleton allows us to stand upright and acts as a foundation for organs and tissues to attach and maintain their location. This is similar to how the wooden frame of a house has room partitions to designate the type of room and floors that furniture can attach to.

The skeletal system and the muscular system are literally interconnected and allow for voluntary movement. Strong connective tissues called **tendons** attach bones to muscles. Most muscles work in opposing pairs and act as levers. For example, flexing the biceps brings the arm bones together and flexing the triceps straightens the elbow. Synovial joints are movable joints, and they are rich with cartilage, connective tissue, and synovial fluid which acts as a lubricant. The majority of joints are synovial joints, and they include hinge joints, like the one at the elbow, which allows for opening and closing.

The vertebrae are cartilaginous joints, which have spaces between them filled with cushion-like discs that act as shock absorbers and glue that holds the vertebrae together. The reason for the tight packing of the vertebrae is to protect the spinal cord inside. This limits their movement, but because there are so many of them, it allows the backbone to be flexible.

Fibrous joints like those in the skull have fibrous tissue between the bones and no cavity between them. These are fixed joints that are immobile.

Compact and Spongy Bone

Osteoclasts, osteoblasts, and osteocytes are the three types of bone cells. Osteoclasts break down old bone, osteoblasts make new bone, and osteocytes are the mature, functional bone cells. Bone is constantly regenerating due to the osteoblasts/osteoclasts that line all types of bones and the blood vessels inside them. The cells all exist within a matrix of collagen fibers that provide resistance to tension and minerals that provide resistance to compression. Because of the collagen and mineral matrix, bones have ample reinforcement to collectively support the entire human body.

Bones can be classified as any of the following:

- Long bones include tube-like rods like the arm and leg bones.
- Short bones are tube-like rods that are smaller than long bones like the fingers and toes.
- Flat bones are thin and flat like the ribs and breastbone.
- Irregular bones like the vertebrae are compact and don't fit into the other categories.

The outer tissue of the bone is surrounded by connective tissue known as **periosteum**. It appears shiny, smooth, and white. It protects the bone, anchors the bone to the connective tissue that surrounds muscles, and links the bone to the circulatory and nervous system. Compact bone is underneath the periosteum and is made of a dense blend of tightly packed osteocytes. It serves as a mineral reservoir of calcium and phosphorous. Compact bones have a **Haversian system**, which is composed of embedded blood vessels, lymph vessels, and nerve bundles that span the interior of the bone from one end to the other. Branching from the central canal to the surface of the bone are the **canals of Volkmann**, which deliver materials to peripheral osteocytes. Concentric circles surround the central Haversian canal, and these **lamellae** have gaps between them called **lacunae** where osteocytes are embedded.

In contrast, spongy bone is very porous and more flexible than compact bone. It is at the ends of long bones and the central part of flat bones. It looks like a honeycomb, and the open spaces are connected by **trabeculae**, which are beams of tissue that add support. Trabeculae add strength without adding mass.

Cartilage is a very flexible connective tissue made of collagen and the flexible elastin. It has no blood vessels and obtains materials via diffusion. It is replaced by bone starting in infancy in a process called **ossification**.

Life and Physical Sciences

Basic Macromolecules in a Biological System

Biological Molecules

All biological material is made up of atoms bonded together to form molecules. These bonds give the molecules their structure and determine the function of the molecule. Many biologically important molecules are polymers, which are chains of a repeated basic unit, called a monomer.

Nucleic Acids

Two types of five-carbon sugars can form nucleotides: ribose and deoxyribose. Ribose is the sugar found in ribonucleic acid (RNA). Deoxyribose is the sugar found in deoxyribonucleic acid (DNA), and it has one less oxygen atom than ribose. The important nitrogenous bases are adenine, guanine, cytosine, thymine, and uracil. Thymine is found only in DNA and uracil is found only in RNA. The structure of these molecules determines their function. DNA and RNA have different properties because ribose and deoxyribose have slightly different structures. DNA is usually found as a double helix because deoxyribose has more flexibility. The specific function of the nucleic acid is also determined by the sequence and properties of the nitrogenous bases. Nitrogenous bases form hydrogen bonds with specific other nitrogenous bases. Adenine interacts with thymine and uracil, and cytosine interacts with guanine; these interactions are called **base pairs**.

This specific pairing allows DNA to serve as the hereditary material for cells because it can be copied accurately, in just one way, so that it is passed down correctly to daughter cells. DNA is also able to serve as hereditary material because the sequence of the nitrogenous bases acts as a code that can be made into all of the proteins needed by the cell. RNA is transcribed from DNA accurately because the nitrogenous bases from RNA interact with those from DNA. Proteins are translated from this RNA using a special type of RNA, called a **transfer RNA**. The transfer RNA has three nucleotides that can bind to the RNA being used to make the protein.

Proteins

Proteins are molecules that consist of carbon, hydrogen, oxygen, nitrogen, and other atoms, and they have a wide array of functions. The monomers that make up proteins are amino acids. All amino acids have the same basic structure. They contain an amine group (-NH), a carboxylic acid group (-COOH), and an R group. The R group, also called the **functional group**, is different in each amino acid.

The functional groups give the different amino acids their unique chemical properties. There are twenty naturally-occurring amino acids that can be divided into groups based on their chemical properties. Glycine, alanine, valine, leucine, isoleucine, methionine, phenylalanine, tryptophan, and proline have nonpolar, hydrophobic functional groups. Serine, threonine, cysteine, tyrosine, asparagine, and glutamine have polar functional groups. Arginine, lysine, and histidine have charged functional groups that are basic, and aspartic acid and glutamic acid have charged functional groups that are acidic.

A **peptide bond** can form between the carboxylic-acid group of one amino acid and the amine group of another amino acid, joining the two amino acids. A long chain of amino acids is called a **polypeptide** or protein. Because there are so many different amino acids and because they can be arranged in an infinite number of combinations, proteins can have very complex structures. There are four levels of protein structure. Primary structure is the linear sequence of the amino acids; it determines the overall structure of the protein and how the functional groups are positioned in relation to each other, as well as how they interact. Secondary structure is the interaction between different atoms in the backbone chain of the protein. The two main types of secondary structure are the alpha helix and the beta sheet. Alpha helices are formed when the N-H of one amino-acid hydrogen bonds with the C=O of an amino acid four amino acids earlier in the chain.

The functional groups of certain amino acids—including methionine, alanine, uncharged leucine, glutamate, and lysine—make the formation of alpha helices more likely. The functional groups of other amino acids, such as proline and glycine, make the formation of alpha helices less likely. Alpha helices are right-handed and have 3.6 residues per turn. Proteins with alpha helices can span the cell membrane and are often involved in DNA binding. Beta sheets are formed when a protein strand is stretched out, allowing for hydrogen bonding with a neighboring strand.

Similar to alpha helices, certain amino acids have an increased propensity to form beta sheets. Tertiary protein structure forms from the interactions between the different functional groups and gives the protein its overall geometric shape. Interactions that are important for tertiary structure include hydrophobic interactions between nonpolar side groups, hydrogen bonding, salt bridges, and disulfide bonds. Quaternary structure is the interaction that occurs between two different polypeptide chains and involves the same properties as tertiary structure. Only proteins that have more than one chain have quaternary structure.

Lipids

Lipids are a very diverse group of molecules that include fats, oils, waxes, and steroids. Since most lipids are primarily nonpolar, they are hydrophobic; however, some lipids have polar regions, making them **amphiphilic**, which means they are both hydrophobic and hydrophilic. Because the structure of different lipids is so diverse, they have a wide range of functions, which include energy storage, signaling, structure, protection, and making up the cell membrane. Triglycerides are one type of biologically-important lipid. They are made up of one molecule of glycerol bonded to three long fatty-acid chains, which are long hydrocarbon chains with a carboxylic acid group.

Many of the properties of triglycerides are determined by the structure of the fatty-acid chains. Saturated triglycerides are made of hydrocarbon chains that only have single bonds between the carbon atoms. Because this structure can pack in more tightly, these triglycerides are often solids at room temperature. Unsaturated triglycerides have double bonds between one or more pairs of carbon atoms. These double bonds create a kink in the fatty-acid chain, which prevents tight packing of the molecules; thus, triglycerides are often liquid at room temperature.

Carbohydrates

Carbohydrates are molecules consisting of carbon, oxygen, and hydrogen atoms and serve many biological functions, such as storing energy and providing structural support. The monomer of a carbohydrate is a sugar **monosaccharide**. Some biologically important sugars include glucose, galactose, and fructose. These monosaccharides are often found as six-membered rings and can join together by a process known as condensation to form a disaccharide, which is a polymer of two sugar units.

Polysaccharides are long chains of monosaccharides. Their main biological functions are to store energy and provide structure. Starch and cellulose are both polymers of glucose made by plants, but their functions are very different because the structure of the polymers are different. Plants make starch to store the energy that is produced from photosynthesis, while cellulose is an important structural component of the cell wall. These two molecules differ in how the molecules of glucose are bonded together. In starch, the bond between the two glucose molecules is a high-energy alpha bond that is easily hydrolyzed, or broken apart. The bonds between the glucose molecules in cellulose are beta bonds. Being stiff, rigid, and hard to hydrolyze, these bonds give the cell wall structural support. Other important carbohydrates include glycogen, which provides energy storage in animal cells, and chitin, which provides structure to arthropod exoskeletons.

Enzymes

Enzymes are biological molecules that accelerate the rate of chemical reactions by lowering the activation energy needed to make the reaction proceed. Although most enzymes can be classified as proteins, some are ribonucleic acid (RNA) molecules. Enzymes function by interacting with a specific substrate in order to create a different molecule, or product. Most reactions in cells need enzymes to make them occur at rates fast enough to sustain life.

Directionality Affects the Structure and Function of a Polymer

The structure of these biological molecules often results in directionality, where one end is different from the other.

In nucleic acids, the five carbons that make up the sugar molecule are numbered. The 5′ carbon is bound to the phosphate group and the 3′ carbon is bound to a hydroxyl group (-OH). A nucleotide strand can only grow by bonding the hydroxyl group to the phosphate group, so one end of nucleic acids is always a 5′ end and the other is always a 3′ end. New nucleotides can only be added to the 3′ end of the nucleic-acid strand. Therefore, replication and transcription only occur in the 5′ to 3′ direction.

Proteins also have directionality. One end of an amino acid contains the NH group and the other end contains the COOH group. This structure gives the amino acids and peptides directionality. A peptide bond can only form between the NH group of one amino acid and the peptide group of another.

Carbohydrates consist of sugars and polymers of sugars. The simplest sugars are monosaccharide, which have the empirical formula of CH_2O. The formula for the monosaccharide glucose, for example, is $C_6H_{12}O_6$. Glucose is an important molecule for cellular respiration, the process of cells extracting energy by breaking bonds through a series of reactions. The individual atoms are then used to rebuild new small molecules. It is important to note that monosaccharides are the smallest functional carbohydrates, so they cannot be hydrolyzed into simpler molecules and still remain carbohydrates.

Polysaccharides are made up of a few hundred to a few thousand monosaccharides linked together. These larger molecules have two major functions. The first is that they can be stored as starches, such as glycogen, and then broken down later for energy. Secondly, they may be used to form strong materials, such as cellulose, which is the firm wall that encloses plant cells, and chitin, the carbohydrate insects use to build exoskeletons.

Changes in Structure Can Cause Changes in Function

Interactions between molecules are the driving force behind biological processes. The structures of the different molecules facilitate their interaction. Biological processes progress because this interaction causes a change in the structure of one or more of the molecules, facilitating interactions with new molecules.

The Effects of Enzyme Shapes

As mentioned, enzymes are proteins that catalyze a reaction, which means they make a reaction happen faster. Almost all biological reactions are catalyzed by enzymes. In order for the enzyme to function, it needs to interact with the substrates, which are the components of the reaction. This process is very precise to ensure that only the correct substrates are involved in the reaction. Therefore, the substrates have a very specific fit into the active site of the enzyme, which is the area that catalyzes the reaction. Once the enzyme and substrate bind to the active site, they form what is called the **enzyme/substrate complex.**

Cofactors and Coenzymes

To function correctly, some enzymes require **cofactors**, such as vitamins, and coenzymes. These molecules affect the shape of the enzyme. When they are not present, the substrate cannot fit into and interact with the enzyme. However, when the cofactors or coenzymes are present, the shape of the enzyme changes, and it can interact with the substrate.

Because this fit needs to be so perfect, it offers a method for regulation of enzyme activity. Other factors or molecules may bind to the enzyme and change its shape. This alters the enzyme's ability to bind to the substrates. These molecules bind to allosteric sites, which are sites that are different from the active site. This binding can either change the shape to allow the substrate to fit, which is called **allosteric activation**, or it can prohibit the substrate from fitting, which is called **allosteric inhibition**.

Prokaryotes and Eukaryotes

Biological molecules interact with each other to form different structures. The basic unit of biological life is the cell. There are two distinct types of cells that make up most living organisms: prokaryotic and eukaryotic. Both types of cells are enclosed by a cell membrane, which is selectively permeable. Selective permeability means essentially that it is a gatekeeper, allowing certain molecules and ions in and out, and keeping unwanted ones at bay, at least until they are ready for use. They both contain ribosomes and DNA. One major difference between these types of cells is that in eukaryotic cells, the cell's DNA is enclosed in a membrane-bound nucleus, whereas in prokaryotic cells, the cell's DNA is in a region—called the **nucleoid**—that is not enclosed by a membrane. Another major difference is that eukaryotic cells contain organelles, while prokaryotic cells do not have organelles.

Prokaryotic cells include bacteria and archaea. They do not have a nucleus or any membrane-bound organelles, are unicellular organisms, and are generally very small in size. Eukaryotic cells include animal, plant, fungus, and protist cells. Fungi are unicellular microorganisms such as yeasts, molds, and mushrooms. Their distinguishing characteristic is the chitin that is in their cell walls. Protists are organisms that are not classified as animals, plants, or fungi; they are unicellular; and they do not form tissues.

Structure and Function of Cell Membranes

All cells are surrounded by a cell membrane that is formed from two layers of phospholipids. **Phospholipids** are two fatty acid chains connected to a glycerol molecule with a phosphate group. The membrane is amphiphilic because the fatty acid chains are hydrophobic and the phosphate group is hydrophilic. This creates a unique environment that protects the cell's inner contents while still allowing material to pass through the membrane. Because the outside of a cell, known as the **extracellular space**, and the inside of a cell, the **intercellular space**, are aqueous, the lipid bilayer forms with the two phospholipid heads facing the outside and the inside of the cell. This allows the phospholipids to interact with water; the fatty acid tails face the middle, so they can interact with each other and avoid water.

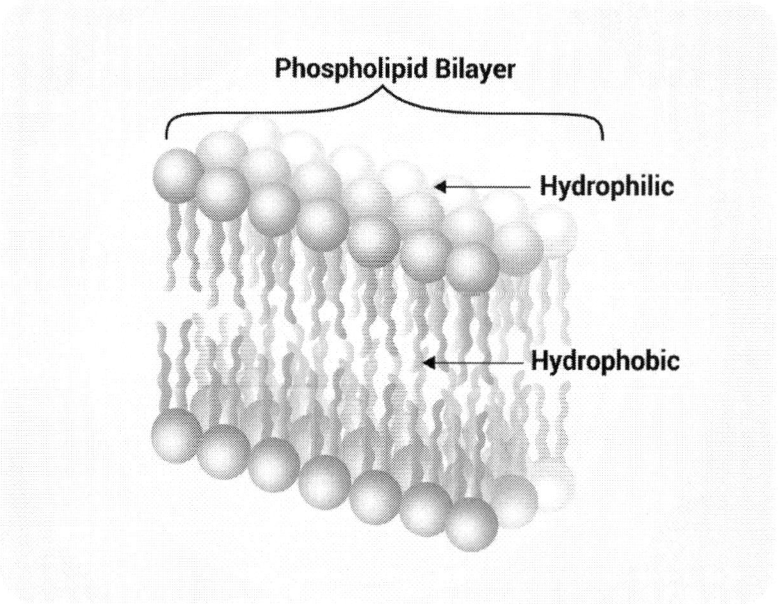

Molecules can pass through the cell membrane by either active or passive transport. **Active transport** requires chemical energy to move molecules in or out of the cell from areas of low concentration to areas of high concentration, or in instances where the molecules would not be able to pass through by themselves, such as with large non-lipid soluble molecules. Ions, amino acids, and complex sugars use active transport mechanisms. Molecules that are soluble in lipids, water, and oxygen use **passive transport** to move in and out of the cell, which means that cellular energy is not required for their movement. Examples of passive transport include diffusion, facilitated diffusion, and osmosis. **Diffusion** is the net movement of particles from an area of high concentration to lower concentration. **Facilitated diffusion** is the movement of molecules through cell membranes with the use of special transport proteins. Finally, **osmosis** is the movement of water molecules across partially permeable membranes.

Cell Organelles Provide Essential Cell Processes

Eukaryotic cells comprise different subcellular components called organelles, whose structure, function, and interactions help the cell undergo cellular processes. Each organelle has an essential function for sustaining the life of the cell.

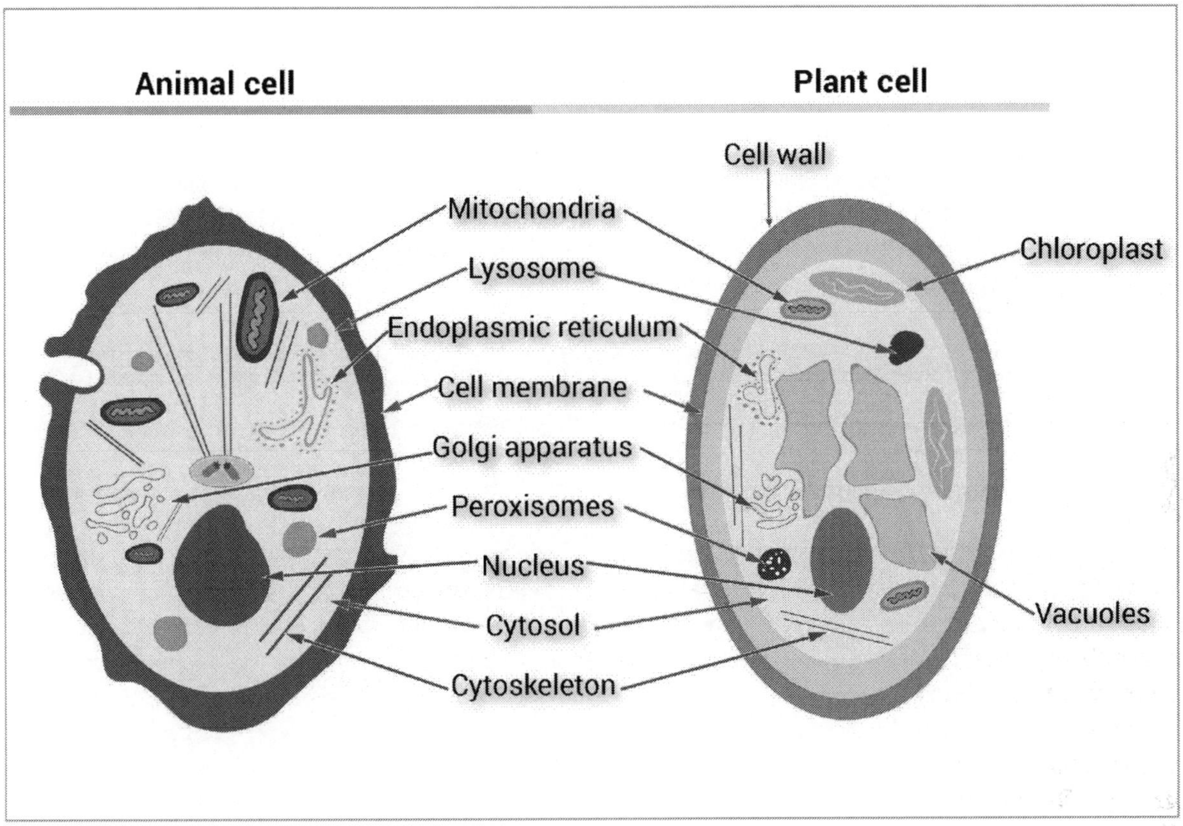

Ribosomes

Ribosomes are made up of ribosomal RNA molecules and a variety of proteins. They are the structures that synthesize proteins. They consist of two subunits, small and large. The ribosomes use messenger RNA as a template for the protein and transfer RNA to bring amino acids to the ribosomes, where they are synthesized into peptide strands using the genetic code provided by the messenger RNA. Most ribosomes are attached to the ER membrane.

Endoplasmic Reticulum (ER)

The ER is a network of membranes contiguous with the outer membrane of the nucleus that forms flat disks. There are two types of ER, rough and smooth. Rough ER is lined with ribosomes and is the location of protein synthesis. This provides a separate compartment for site-specific protein synthesis and is important for the intracellular transport of proteins. Smooth ER does not contain ribosomes and is the location of lipid synthesis.

Golgi Apparatus

The Golgi apparatus is the site where proteins are modified; it is involved in the transport of proteins, lipids, and carbohydrates within the cell. The Golgi apparatus is made up of flat layers of membranes called **cisternae**. Material is transported in transfer vesicles from the ER to the cis region of the Golgi apparatus. From there, the material moves through the medial region, is sometimes modified, and then leaves through the trans region of the ER in a secretory vesicle.

Mitochondria

The mitochondrion is the primary site of respiration and adenosine triphosphate (ATP) synthesis inside the cell. Mitochondria have two lipid bilayers that create the intermembrane space, which is the space between the two membranes, and the matrix, which is the space inside the inner membrane. While the outer membrane is smooth, the inner membrane is folded and forms cristae. The outer membrane is permeable to small molecules. The cristae contain many proteins involved in ATP synthesis. In the mitochondria, the products of glycolysis are further oxidized during the citric-acid cycle and the electron-transport chain. The two-layer structure of the mitochondria allows for the buildup of H+ ions produced during the electron-transport chain in the intermembrane space, creating a proton gradient and an energy potential. This gradient drives the formation of ATP. Mitochondria have their own DNA and are capable of replication.

Lysosomes

Lysosomes are membrane-bound organelles that perform intracellular digestion. These organelles have a low pH and contain digestive enzymes. The structure of lysosomes allows for the recycling of cellular material while protecting the rest of the cell from harmful pH and enzymes.

Vacuoles

Vacuoles are membrane-bound organelles found primarily in plant and fungi cells, but also in some animal cells. Vacuoles are filled with water and some enzymes and are important for intracellular digestion and waste removal. The membrane-bound nature of the vacuole allows for the storage of harmful material and poisonous substances. The pressure from the water inside the vacuole also contributes to the structure of plant cells.

Chloroplasts

Chloroplasts are organelles found primarily in plants and are the site of photosynthesis. They have a double membrane and also contain membrane-bound **thylakoids**, or discs, that are organized into **grana**, or stacks. Chlorophyll is present in the thylakoids, and the light stage of photosynthesis, which includes the production of ATP and $NADPH_2$, occurs there. Chlorophyll is green and traps the light energy necessary for photosynthesis. Chloroplasts also contain stroma, which are the site of the dark reaction stage of photosynthesis, during which sugar is made. The membrane structure of chloroplasts allows for the compartmentalization of the light and dark stages of photosynthesis.

Chromosomes, Genes, and DNA

The Source of Heritable Information

DNA was not always believed to be the cell's genetic material. Significant scientific contributions included:

<u>Griffith, 1920s</u>

Experimental Design & Results

- Infected mice with harmful and benign bacteria.
- If harmful bacteria were killed and injected, no effect
- If harmful bacteria were killed and mixed with living benign bacteria, some part of the harmful bacteria transformed the harmless bacteria

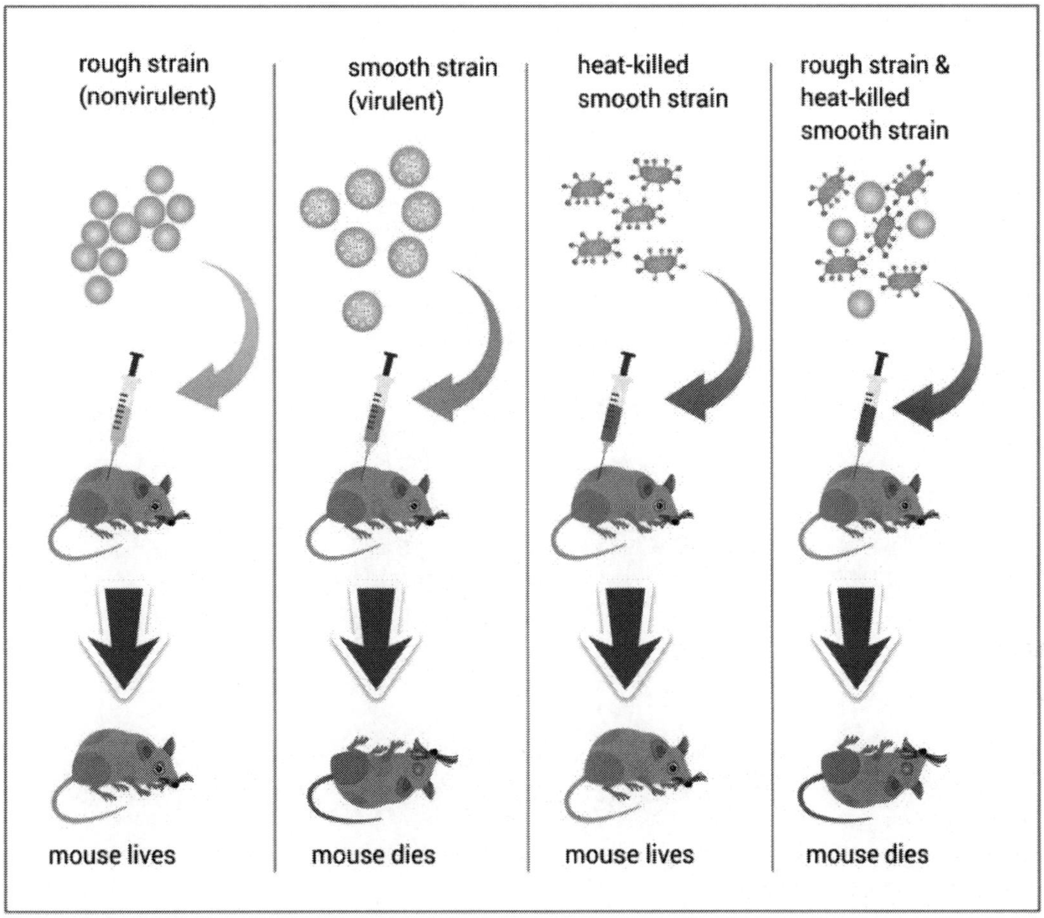

Discovery & Significance

- Something in a cell is capable of passing instructions
- Coined the term "transformation" or the process when a cell picks up foreign DNA and then changes phenotype

Avery, 1930s & 1940s

Experimental Design & Results

- Did the same experiment expect he used a substance to deactivate either DNA, RNA, or protein
- He found that only when DNA was active did the harmless bacteria transform the cell, indicating that DNA, not RNA or protein, is the genetic material.

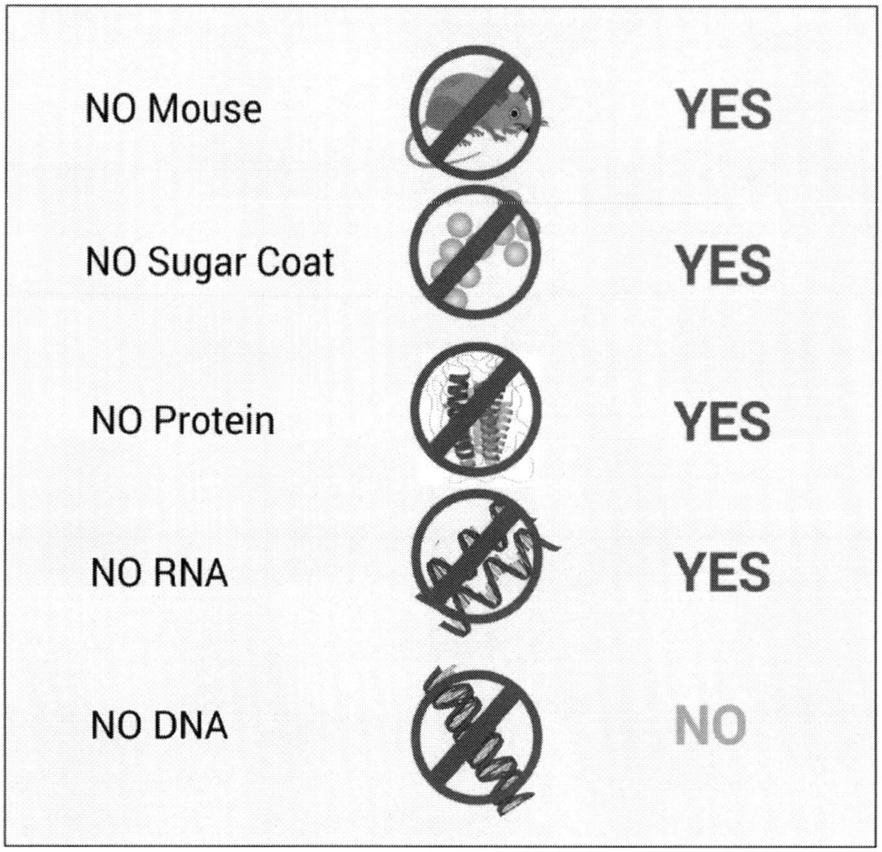

Discovery & Significance

- Skeptics dismissed the results due to the belief that bacteria and eukaryotes have different processes.

Hershey & Chase, 1950s

Experimental Design & Results

- Different radiolabels were use on proteins and DNA in phages that infected bacterial cells.
- It was observed that the DNA label was inside the bacteria while the protein label was in the supernatant.

Discovery & Significance

- DNA is the component that enters cells, therefore DNA is the genetic material.

Rosalind Franklin and Watson & Crick 1940s & 1950s

Experimental Design & Results

- Used X-ray crystallography to photograph DNA
- Used Franklin's data to finalize and publish the double-helix model of DNA

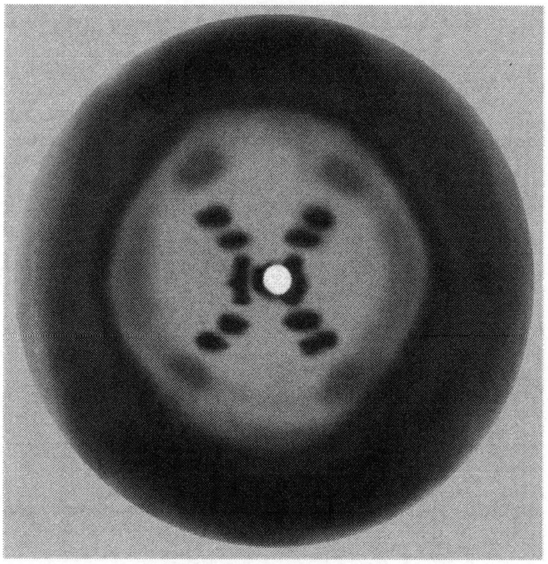

Discovery & Significance

- Discovered that the coiled, double stranded molecule was arranged in an anti-parallel manner
- DNA had a sugar-phosphate backbone held together by nitrogen bases

Meselson & Stahl, 1950s

Experimental Design & Results

- Radio-labelled DNA with a heavy isotope, then placed it in a solution where it replicated with a lighter labeled isotope
- If DNA replication is conservative, then the original heavy DNA would still be present (not observed).
- If DNA replication is dispersive, there would only be one thick medium band (not observed).
- Semi-conservative replication would result in a thin medium band and more and more light bands since the DNA was only replicating with the light isotope (observed).

Discovery & Significance

- DNA replication is semi-conservative, meaning that when the strands are separated, new DNA is built from each parent strand. Each new DNA molecule has one old strand and one new strand.

Structure and Function of DNA

Deoxyribonucleic acid (DNA) is life's instruction manual. It is double stranded and directional, meaning it can only be transcribed and replicated from the *3' end* to the *5' end*.

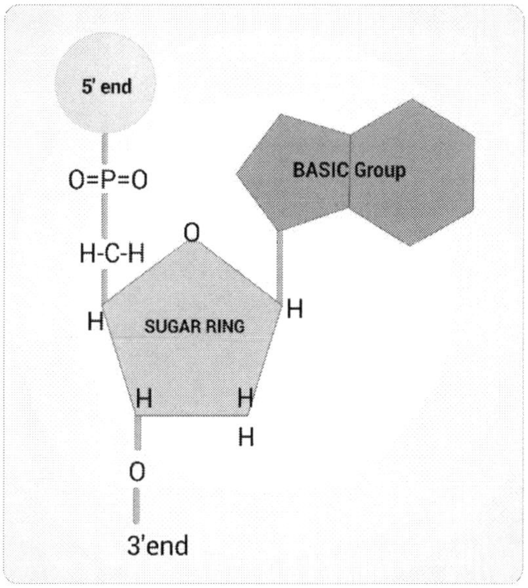

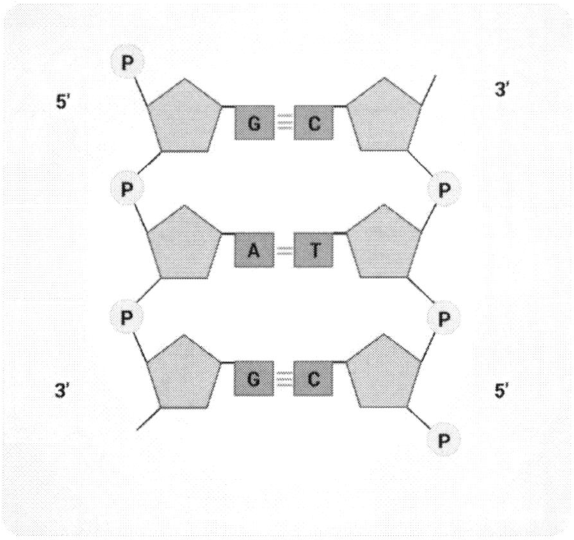

DNA as a Monomer	DNA as a Polymer
A nucleotide is composed of a five-carbon sugar with a phosphate group off of the 5th carbon and a nitrogenous base off of the 1st carbon. DNA and RNA are different because DNA contains deoxyribose sugar while RNA contains ribose sugar. Also, the nitrogenous base thymine in DNA is replaced by uracil in RNA.	The two strands are **antiparallel**, meaning they are read in opposite directions. The bases guanine and cytosine are complementary and held together by three hydrogen bonds, and the bases adenine and thymine are complementary and held together by two hydrogen bonds. Weak hydrogen bonding between bases allows DNA to be opened easily for transcription and replication.

Nitrogen bases come in two varieties:

- One-Ringed Pyrimidines: Cytosine and Thymine (DNA)/Uracil (RNA)
- Two-Ringed Purines: Adenine and Guanine

<u>Purines and Pyrimidines</u>
The five bases in DNA and RNA can be categorized as either pyrimidine or purine according to their structure. The pyrimidine bases include cytosine, thymine, and uracil. They are six-sided and have a single ring shape. The purine bases are adenine and guanine, which consist of two attached rings. One ring has five sides and the other has six. When combined with a sugar, any of the five bases become nucleosides. Nucleosides formed from purine bases end in "osine" and those formed from pyrimidine bases end in "idine." Adenosine and thymidine are examples of nucleosides. Bases are the most basic components, followed by nucleosides, nucleotides, and then DNA or RNA.

Structure and Function of RNA

RNA is short for ribonucleic acid, which is a type of molecule that consists of a long chain (polymer) of nucleotide units. RNA and DNA differ in terms of structure and function. RNA has a different sugar than DNA. It has ribose rather than deoxyribose sugar. The RNA nitrogenous bases are adenine (A), guanine (G), cytosine (C), and uracil (U). Uracil is found only in RNA and thymine is found only in DNA. RNA consists of a single strand and DNA has two strands. If straightened out, DNA has two side rails. RNA only has one "backbone," or strand of sugar and phosphate group components. RNA uses the fully hydroxylated sugar pentose, which includes an extra oxygen compared to deoxyribose, which is the sugar used by DNA. RNA supports the functions carried out by DNA. It aids in gene expression, replication, and transportation.

RNA acts as a helper to DNA and carries out a number of other functions. Types of RNA include ribosomal RNA (rRNA), transfer RNA (tRNA), and messenger RNA (mRNA). Viruses can use RNA to carry their genetic material to DNA. Ribosomal RNA is not believed to have changed much over time. For this reason, it can be used to study relationships in organisms. Messenger RNA carries a copy of a strand of DNA and transports it from the nucleus to the cytoplasm. DNA unwinds itself and serves as a template while RNA is being assembled. The DNA molecules are copied to RNA. **Translation** is the process whereby ribosomes use transcribed RNA to put together the needed protein. Transfer RNA is a molecule that helps in the translation process, and is found in the cytoplasm. Ribosomal RNA is in the ribosomes.

Transcription and Translation

<u>Transcription</u>

Transcription is the process by which a segment of DNA is copied onto a working blueprint called RNA. Each gene has a special region called a promoter that guides the beginning of the transcription process. RNA polymerase unwinds the DNA at the promoter of the needed gene. After the DNA is unwound, one strand or template is copied by the RNA polymerase by adding the complementary nucleotides, G with C, C with G, T with A, and A with U. Then, the sugar phosphate backbone forms with the aid of RNA polymerase. Finally, the hydrogen bonds joining the strands of DNA and RNA together are broken. This forms a single strand of messenger RNA or mRNA.

Codons are groups of three nucleotides on the messenger RNA, and can be visualized as three rungs of a ladder. A codon has the code for a single amino acid. There are 64 codons but 20 amino acids. More than one combination, or triplet, can be used to synthesize the necessary amino acids. For example, AAA (adenine-adenine-adenine) or AAG (adenine-adenine-guanine) can serve as codons for lysine. These groups of three occur in strings, and might be thought of as frames. For example, AAAUCUUCGU, if read in groups of three from the beginning, would be AAA, UCU, UCG, which are codons for lysine, serine, and serine, respectively. If the same sequence was read in groups of three starting from the second position, the groups would be AAU (asparagine), CUU (proline), and so on. The resulting amino acids would be completely different. For this reason, there are start and stop codons that indicate the beginning and ending of a sequence (or frame). AUG (methionine) is the start codon. UAA, UGA, and UAG, also known as ocher, opal, and amber, respectively, are stop codons.

Prokaryotic transcription consists of the following three stages:

1. Initiation: RNA polymerase binds to the start site on the RNA template strand, which is upstream in the promotor region of the gene.
2. Elongation: Synthesis of the new complementary RNA strand begins by the action of RNA polymerase as it works its way down the gene, growing the new strand.
3. Termination: The new RNA chain is completed and the RNA polymerase is released at the stop point, which is downstream on the gene in the terminator region.

Eukaryotic transcription is more complicated:

1. Eukaryotic Initiation:: Eukaryotes contain advanced **promoters**, which are DNA sequences upstream of genes that recruit transcription factors, proteins that facilitate or block RNA polymerase binding. The TATA box is where RNA polymerase II binds.
2. Eukaryotic Elongation:: This is similar to prokaryotic elongation in that it generates a 5' to 3' complementary RNA transcript.
3. Eukaryotic Termination: A polyadenylation signal (DNA sequence AAUAAA) causes RNA polymerase to release several bases downstream. The RNA transcript is then processed. Upstream of the promoter, past a 5' untranslated region, a "cap" is added (like a telomere, but only a few repeated Guanines). Downstream of the polyadenylation signal, a poly(A) tail is added. In addition to changes at the end of each transcript, areas within the transcript are further processed through splicing. **Spliceosomes** (large complexes of RNA and proteins) remove intermittent noncoding regions called **introns**, and the **exons** (coding sequences) are joined together. The introns are spliced out so that only the exons are part of the final transcript. Alternative splicing (when one transcript is spliced in different ways and creates different proteins) makes this process even more complicated.

Transcription Regulation

Eukaryotic transcription initiation involves transcription factors. Different transcription factors are present in different cell types, and certain sequences of DNA enhance protein binding. This results in an altered DNA spatial arrangement and exposure to RNA polymerase. Some transcription factors facilitate transcription while some block it. The presence or absence of these proteins in certain cells is one way gene production is regulated.

Another way transcription is regulated in eukaryotes is by gene accessibility. DNA is wrapped around histone proteins in complexes called **nucleosomes.** These nucleosomes coil and supercoil, which make portions of the genome inaccessible. Modification of histone tails in nucleosomes open and close regions of DNA, making them either more or less available for the protein binding of transcription factors and RNA polymerase. When the chromatin is in its closed, coiled conformation, called **heterochromatin,** transcription is repressed because the genes are inaccessible. In contrast, in the **euchromatin** formation, the chromatin is open and uncoiled, which allows RNA polymerase to access the genes, which activates transcription.

Essentially, open chromatin (*euchromatin*) has acetylated histone tails with few methyl groups. Heterochromatin is the opposite, with histone tails being highly methylated and deacetylated so that DNA is closed and condensed.

MicroRNAs (miRNAs) and small interfering RNAs (siRNAs) can also regulate gene production. This is done by degrading certain transcripts or blocking their translation, and sometimes even altering chromatin structure.

It should be noted that there is far more DNA functionality than gene expression. DNA also contains promoters and enhancers to regulate gene expression, centromeres, telomeres, transposable elements, and other sequences.

Transposable elements are literally "jumping genes" that relocate within an organism's genome. Prokaryotes and eukaryotes contain **transposons.** This supports the idea that they have a significant effect on biodiversity and arose through common ancestry among all living organisms. DNA sequences prone to transposon trespassing are predictable, repeated bases that can extend from 300 to 30,000 nucleotides.

Within an organism, transposons relocate and might leave the targeted segment behind. A transposon can make a copy of the DNA or simply cut out the sequence. Either way, transposase is a critical enzyme involved in the process.

Retrotransposons are elements that copy the targeted segment into an RNA transcript. The enzyme retrotransposase is then used to copy the intermediate back into DNA before inserting it, thus copying part of the genome and moving it elsewhere.

It is believed that all human ancestors once had brown eyes until a single individual had a mutation which coded for blue eyes. Once that allele entered the gene pool, it was passed to generations of offspring. Today it is fairly common to see a blue-eyed individual. Thus, changes in alleles can change phenotypes.

Translation

Translation is the process of generating protein from RNA. Translation in prokaryotes is far simpler than in eukaryotes. They only have one circular chromosome (thus, far fewer genes), and they do not have DNA within a nucleus. They also do not process transcripts. Translation can even occur simultaneously with transcription on the same piece of RNA. In fact, many different ribosomes can be working on the same transcript at the same time, thus creating a structure called a **polyribosome.**

The ribosomes of bacteria and eukaryotes are similar. A ribosome has two subunits, both made of ribosomal RNA (*rRNA*, which is made by RNA polymerase I in eukaryotes). In eukaryotes, the small subunit and the large subunit assemble in conjunction with the AUG start codon in the mRNA message (DNA is read in three letter "words" called codons).

Before translation can start, a transfer RNA molecule (*tRNA*, which is made by RNA polymerase III) must join the ribosomal complex. tRNAs contain anticodons that are complementary to specific codons on the RNA transcript. On one side, their highly specific anticodon binds to corresponding codons in the mRNA. On the other side, tRNA molecules carry specific amino acids, the monomer of proteins. Once a tRNA molecule has released its amino acid, an enzyme amino-acyl-tRNA synthetase will join a free-floating tRNA and its corresponding amino acid. As a result, the tRNA can continue to deliver fresh amino acids to the ribosomes.

The point of translation is that, every time a tRNA molecule drops off an amino acid, it contributes to an emerging protein. Only when a stop codon is reached will the ribosome disassemble, thus releasing the assembled protein.

Second Base in Codon

First Base in Codon	U	C	A	G	Third Base in Codon
U	UUU Phe UUC Phe UUA Leu UUG Leu	UCU Ser UCC Ser UCA Ser UCG Ser	UAU Tyr UAC Tyr UAA Stop UAG Stop	UGU Cys UGC Cys UGA Stop UGG Trp	U C A G
C	CUU Leu CUC Leu CUA Leu CUG Leu	CCU Pro CCC Pro CCA Pro CCG Pro	CAU His CAC His CAA Gln CAG Gln	CGU Arg CGC Arg CGA Arg CGG Arg	U C A G
A	AUU Ile AUC Ile AUA Ile AUG Met or Start	ACU Thr ACC Thr ACA Thr ACG Thr	AAU Asn AAC Asn AAA Lys AAG Lys	AGU Ser AGC Ser AGA Arg AGG Arg	U C A G
G	GUU Val GUC Val GUA Val GUG Val	GCU Ala GCC Ala GCA Ala GCG Ala	GAU Asp GAC Asp GAA Glu GAG Glu	GGU Gly GGC Gly GGA Gly GGG Gly	U C A G

Note in the chart above that each codon codes for a specific amino acid, even specifically coding for stop codons. There are many codon combinations, but only 20 amino acids. The redundancy in codon/amino acid pairs is due to the third base—or "**wobble**" position—of tRNA. The first two bases bind so strongly that sometimes the third base does not play much of an active role in connecting the codon and anticodon.

The ribosome has three tRNA landing spots. The first tRNA binds to the start codon in the "P" site, but every tRNA that follows lands at the "A" site. The growing amino acid chain is added to the amino acid in the A site (connecting via a peptide bond). The "E" site is where the naked tRNA exits after it has removed its amino acid chain.

Once a stop codon is reached, the amino acid chain leaves through an exit tunnel. At this point, it is an immature protein that is not properly folded and might be sent to the Golgi for modification.

<u>Regulation of Expression</u>

Transcriptional regulation in prokaryotes occurs via operons. An **operon** is a DNA sequence comprised of related genes which are clustered behind a single promoter that regulates them. Within the promoter is an operator, which is like an on/off switch.

Cell Cycle in Eukaryotes

The eukaryotic cell cycle involves four phases: growth (G_1), replication of DNA (S), preparation for cell division (G_2), and cell division (M). The three phases preceding cell division are collectively called **interphase.**

S Phase: DNA Replication

A cell must duplicate its DNA prior to cell division so that each daughter cell receives the full set of genetic instructions. To do so, cells use many enzymes. Helicase binds to an origin of replication, separates the two DNA strands, and forms a replication bubble. It then travels along the double-stranded DNA, unzipping it. Topoisomerase is an enzyme downstream of helicase that prevents supercoiling of the DNA. It does this by snipping the DNA, swiveling around it, and then pasting the strands back together. Single-stranded binding proteins hold the DNA strands apart at the replication fork.

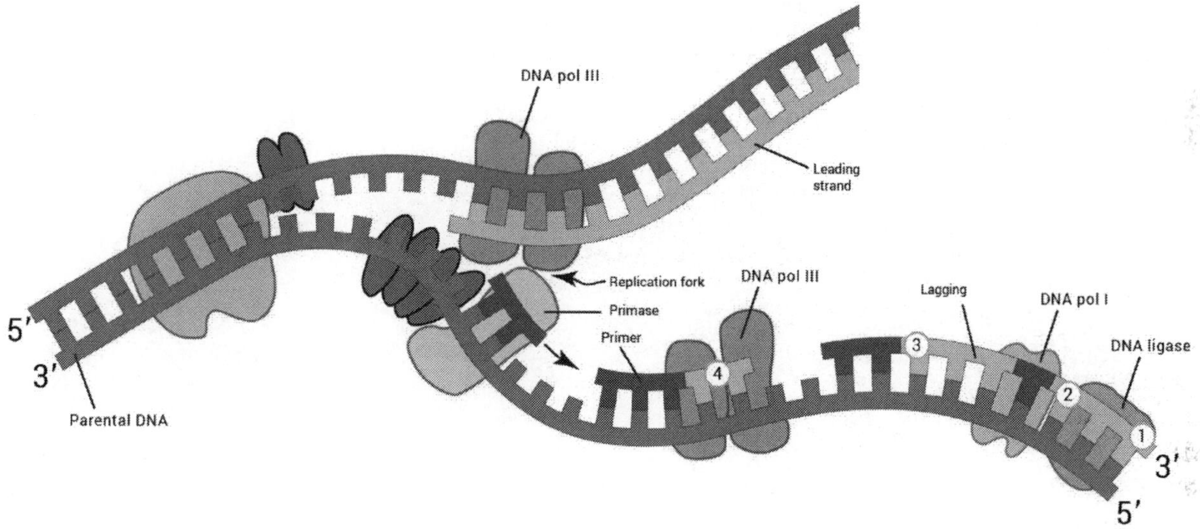

The two DNA strands are copied differently.

Leading Strand – Continuous	Lagging Strand – Interrupted
In this strand, DNA is replicated continuously as the helicase travels. DNA polymerase III is the elongating enzyme that copies the DNA in a complementary fashion. Before DNA polymerase III can attach to the DNA and begin, an enzyme called primase must insert a small RNA primer for the polymerase to bind to because DNA polymerase III is unable to bind directly to DNA.	This synthesis is similar to the leading strand in that primase deposits a primer for DNA polymerase III binding. However, because DNA cannot be created in the 3' to 5' direction, DNA polymerase III replicates DNA in small 5' to 3' chunks called **Okazaki fragments.** The pieces are eventually connected after DNA polymerase I replaces the RNA primer with DNA. An enzyme called ligase glues the small fragments together.

The cell has several levels to correct mistakes during replication:

- DNA polymerase can proofread as it goes.
- Other enzymes can mismatch repair if polymerase fails to recognize mutations.
- Nucleotide excision repair is a higher-tier corrective mechanism. It involves an enzyme called **nuclease** snipping out the mutation, followed by ligase gluing in the correct sequence.

Telomeres are non-coding, repeating sequences of DNA at the ends of each chromosome. The ends of the chromosome lack a 3′ end for DNA polymerase to attach to and replace the RNA primer. As a result, DNA polymerase is unable to bind and, at each division, a little bit of the ends of each chromosome are un-replicated and lost. When the telomeres shorten to nonexistence (at **senescence**), it's a signal to the cell to end its cell cycling days and go through the process of **apoptosis** or intentional cell death. This could have evolved as a natural defense against cancer since older cells, which are more prone to mutations, are killed off before they become dangerous.

Germ cells have the enzyme telomerase that elongates telomeres. This is important for reproductive purposes so that gametes (egg and sperm cells) can continue to have full DNA instructions.

M Phase

Following DNA synthesis is the G2 phase, where the cell assembles the machinery necessary for cell division. Cell division then occurs in several stages:

PHASE	PHASE EVENTS	ANIMAL CELL DIAGRAM	PLANT CELL DIAGRAM
Prophase	Nucleus disappears and DNA condenses into chromosomes. DNA is already wrapped around histone proteins, and it continues to supercoil until it looks like the letter X. Sister chromatids on either side of the X are identical.	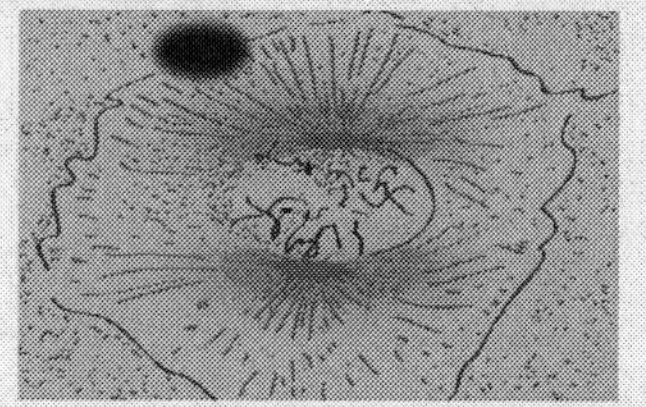	
Metaphase	Chromosomes line up in the cell's center. Kinetochore microtubules extend from animal centrosomes that contain centrioles (organizing centers) on either side of the cell and attach to the centromeres (repeating sequences in the middle of the chromosomes). Nonkinetochore proteins elongate animal cells. This massive protein orchestra is collectively called the spindle apparatus. Plants lack centrioles but have microtubule organizing centers.	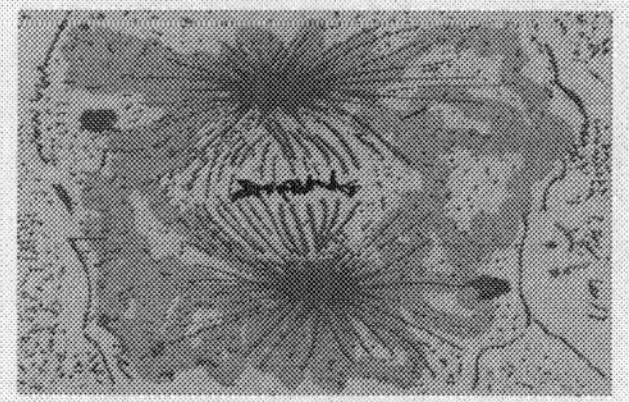	
Anaphase	Kinetochore microtubules shorten/are pulled in and sister chromatids are separated and move to each daughter cell.		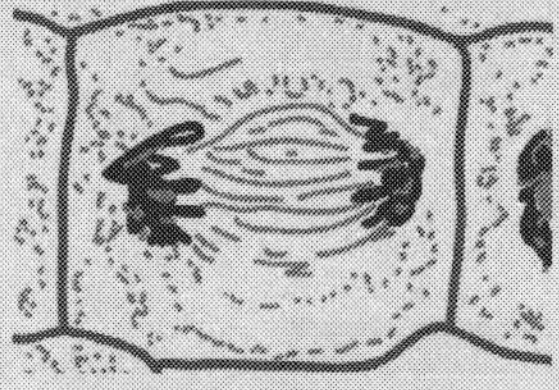
Telophase and Cytokinesis	Nuclei reform and chromosomes decondense within them. **Animals:** actin and myosin microfilaments pinch off the cytoplasm at the cleavage furrow to form two new cells **Plants:** cell plate (new cell wall) forms between daughter cells and extends to divide into two new cells	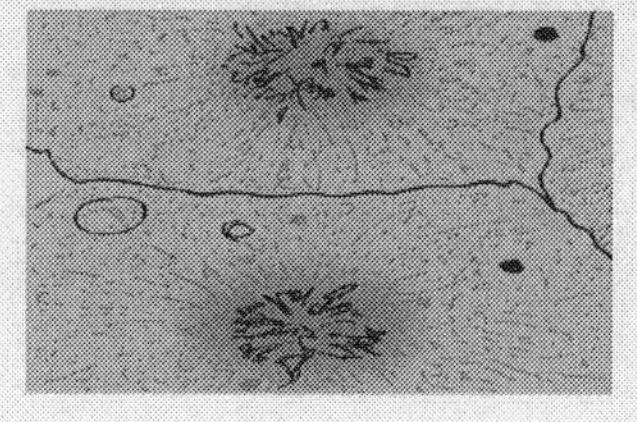	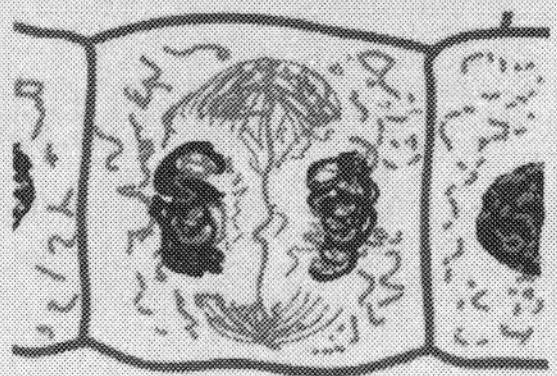

Mitosis occurs for many reasons:

- Development and growth of an organism
- Differentiation and specialization in multicellular organisms
- Replacement of cells with damage or rapid turnover

Cancer cells are the result of inappropriate cell division and occur when cells are unresponsive to checkpoint regulation and growth factor signals. Normal cells are density dependent and anchorage dependent, but cancer cells lose these properties.

Mitosis has several checkpoints. If a cell exits the cell cycle, it enters a phase called G0 that consists of non-dividing cells like neurons. G1 checkpoints prepare and commit cells to entering the cell cycle. S phase proofreads and corrects mistakes. There is also a G2 checkpoint: as a cell progresses through G1, S, and G2, the cyclin protein accumulates. When it becomes abundant, it binds with a cyclin dependent kinase (cdk) to form **Maturation Promoting Factor (MPF),** an activating protein complex that facilitates mitosis. As mitosis is completed, the cyclin is degraded, the MPF complex disassembles, and the cell cycle begins once again.

Meiosis

Meiosis is similar to mitosis because it involves cellular division. However, while mitosis involves the division of somatic (body) cells, meiosis is specifically the production of gametes (egg and sperm). In mitosis, one parent cell splits once into two genetically-identical and diploid daughter cells, while in meiosis, one germ cell splits twice into four genetically different, haploid daughter cells. The two divisions of meiosis (meiosis I and meiosis II) are critical because, when a sperm fertilizes an egg to create the first cell of a new organism (zygote), the zygote must have two sets of chromosomes—not four—to be viable.

The image below shows the steps of meiosis I. These steps look the same as those of mitosis except that the homologous chromosomes line up in the middle of meiosis I. This formation of homologous chromosome pairs is called a **tetrad.**

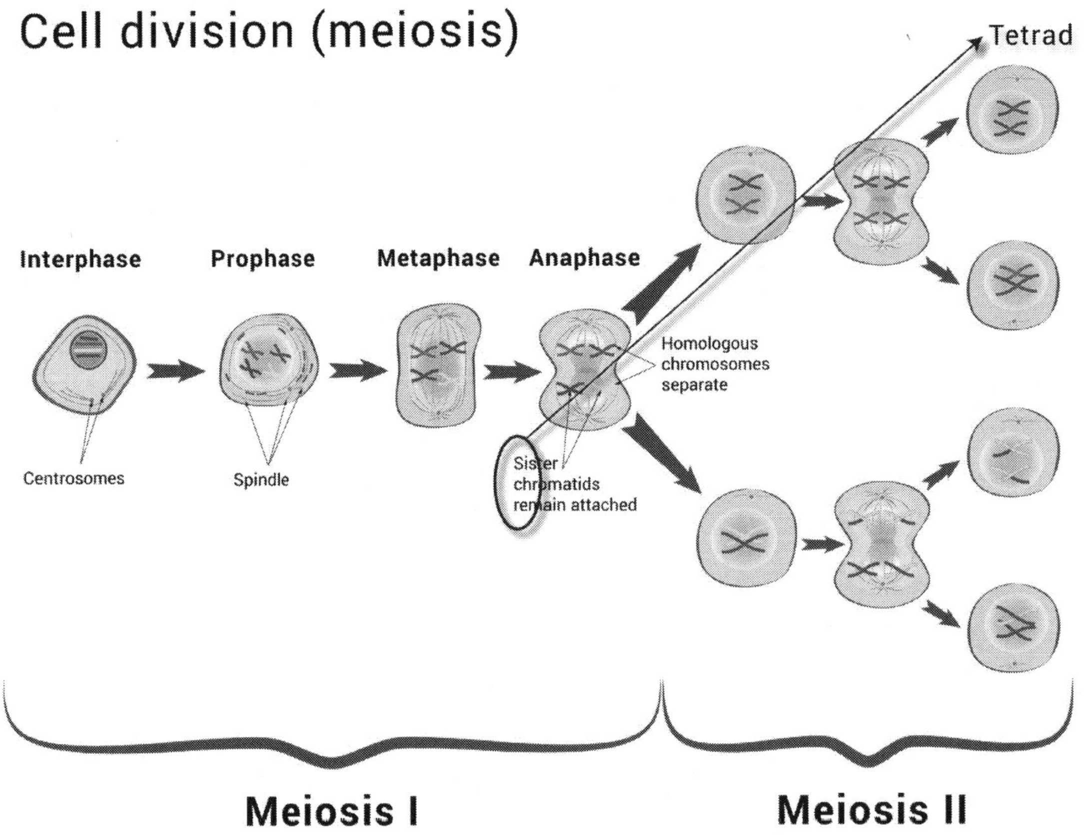

In mitosis, homologous chromosomes line up single file in metaphase. On the other hand, tetrads are paired chromosomes and, in meiosis I, go through a process called **crossing over/recombination** where they exchange DNA. The tetrad is called the **synaptonemal complex**, and a **chiasmata** is where the crossing over occurs (there are 1-3 crossing over events per tetrad). Crossing over makes gametes unique because genetic recombination events are random and unpredictable.

After meiosis I, the homologous chromosomes separate, and the two daughter cells are **haploid,** which means they contain half the normal number of chromosomes. There is no interphase between meiosis I and meiosis II, so the DNA is not replicated. In anaphase II of meiosis, sister chromatids separate and, by the end, there are four unique, haploid daughter cells.

Although homologous chromosomes code for the same genes, there is diversity. This is because each gene has two or more alleles. Alleles are different forms of genes, and the phenotype is dependent on the genotype.

A process called crossing over occurs, which makes the daughter cells genetically different. If chromosomes didn't cross over and rearrange genes, siblings could be identical clones. There would be no genetic variation, which is a critical factor in the evolution of organisms.

Variations are Introduced by the Imperfect Nature of DNA Replication and Repair

Despite replication's proofreading, mutations do happen. Mutations can be either point mutations or chromosomal mutations, and both increase genetic variation.

<u>Point Mutations</u>
Point mutations involve a change in DNA sequence and are due to an addition, deletion, inversion, or substitution error.

Insertion and deletion mutations cause frame-shift mutations. This means that every following codon will be read incorrectly, massively changing the primary structure of the protein. Due to the redundancy of codon/amino acid pairing resulting from the wobble position, some substitution mutations cause no change in the protein sequence. The result is something called a **silent mutation**. All point mutations (including substitution and inversion) can cause gene malfunction.

<u>Chromosomal Mutations</u>
Sometimes there are chromosomal mutations in DNA replication and mitosis. The following types of mutations can occur:

- Deletion: A section of a chromosome is removed.
- Duplication: A section of a chromosome is repeated.
- Inversion: A chromosome is rearranged within itself.
- Translocation: Chromosome pieces mix or combine with other chromosomes.

Not only can mutations lead to changes in an individual's phenotype, but they can also contribute to reproductive isolation and speciation which have effects on a much larger scale.

In metaphase I in meiosis, homologous chromosomes align in the center of the cell. The law of independent assortment states that it is random whether a mother's or father's chromosomes are on the left or the right. Since humans have 23 chromosomes, there are 2^{23} (over 8 million) possible variations. Independent assortment, random fertilization, and recombination contribute greatly to genetic diversity.

Gene Linkage

A man named Thomas Hunt Morgan did extensive genetic research using drosophila (flies), which are an effective test subject because they only have four chromosomes. Using the phenotypes of body color and eye color (grey body and red eyes are dominant), below is a dihybrid cross illustrating the expected phenotypic ratios.

ggrr x GgRr

	gr	gr	gr	gr
GR	GgRr	GgRr	GgRr	GgRr
Gr	Ggrr	Ggrr	Ggrr	Ggrr
gR	ggRr	ggRr	ggRr	ggRr
gr	ggrr	ggrr	ggrr	ggrr

Phenotypic ratio: 4 : 4 : 4 : 4

4	4	4	4
gray body red eyes	gray body purple eyes	black body red eyes	black body purple eyes

However, what Morgan actually observed was not consistent with the expected results. He observed that the cross ggrr x GgRr had a highly skewed phenotypic ratio. Today we understand that this occurs because the two genes are located on the same chromosome and, as such, are more likely to be passed together. Genes close to each other are "linked" and infrequently separate in recombination.

There is a mathematical formula used to find recombination frequency and develop a linkage map illustrating the genes' relative locations on a chromosome. Here is that formula:

$$\frac{Number\ of\ recombinants}{Total\ offspring} \times 100 = Recombination\ frequency$$

Given the calculation results, a linkage map can be constructed to show the relative location since the closer the genes the lower the recombination frequency.

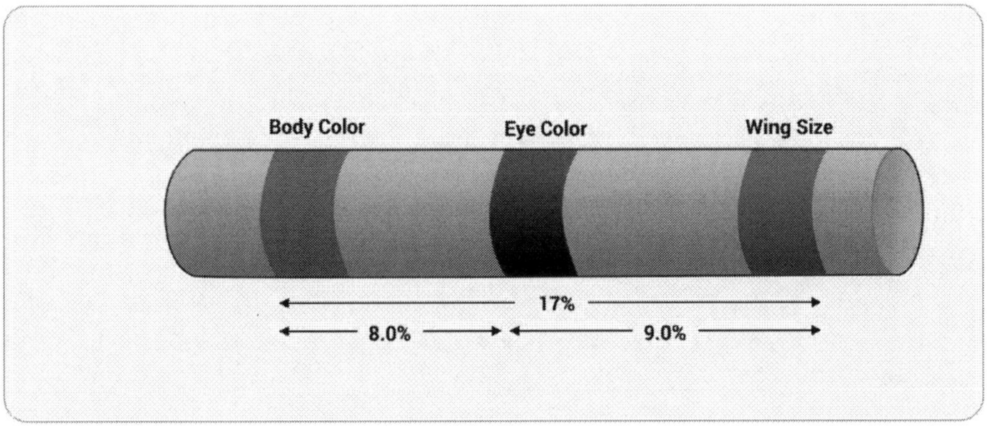

Bacterial Genetic Variation

Occasionally bacteria do have DNA mutations, but they don't harbor chromosomal mutations because they only have one large, central chromosome. They don't perform recombination, instead reproducing asexually via binary fission. However, they can still have genetic variation through the following processes:

1. Transformation
2. Transduction
3. Conjugation
4. Mutation
5. Transposable element relocation

Transformation	Transduction	Conjugation
Exogenous snippets of DNA enter bacterial cells.	Bacteriophages (viruses that infect bacteria) introduce foreign DNA into a bacterial cell	One bacterial cell extends a pili into another and releases DNA.

In conjugation, one bacterial cell is the giver (F^+) while the other is the receiver (F^-). After the pilus extends and attaches to the recipient, it inserts in F factor, a plasmid containing the new allele. Sometimes, R plasmids, which contain and spread antibiotic-resistant genes, are transferred.

Viral Replication

Viruses can induce genetic variation in bacteria via transduction. Viruses (bacteriophages or simply phages) infect bacteria via the lytic or lysogenic cycle. Both processes begin with the bacteriophage injecting its viral genome into a host cell.

Lytic Cycle
This cycle occurs when the viral DNA immediately takes over the host cell, tricking it into copying and translating it. The host cell has no way of telling the difference between the viral and self DNA. By the end of the lytic cycle, hundreds of viruses have been produced and released from the bacterial cell when it "lyses" or bursts.

Lysogenic Cycle
Once the viral DNA enters, it doesn't immediately direct cellular activities. Instead, it embeds itself into the host cell DNA and is called a prophage. This prophage "hides" in the host cell DNA and, as the bacteria replicates via binary fission, it passes its virulent genes on to future generations of cells. Eventually, a signal will activate the viral DNA. When this occurs, the prophage will direct the cell to initiate the lytic cycle, making its host cell a virus-making factory.

Bacteriophages consist of DNA surrounded by a protein coat called a **capsid.** However, they aren't alive because they don't have ribosomes; thus, can't grow or reproduce on their own.

Like bacteriophages, animal viruses contain nucleic acid surrounded by a capsid. They often contain a viral envelope or an outer membrane. Animal viruses can contain DNA as their genetic material, while some contain RNA instead. Those containing RNA may also carry an enzyme called reverse transcriptase, which allows their transcript to be copied into DNA in the host cell's cytoplasm.

Like all viruses, HIV has specificity as to which cells it can infect. This is due to cell recognition. HIV contains certain marker proteins or antigens which are only recognized by a subset of cells, primarily helper T cells. These helper T cells are activators of the entire specific immune response that stimulates antibody production. Should the helper T cell count drop, an individual develops Acquired Immune Deficiency Syndrome (AIDS), a devastating disease that destroys the immune system. If enough T cells are destroyed, AIDS is fatal.

The table below briefly describes the steps in which the HIV retrovirus infects a helper T cell:

Step	Explanation
1	Antigens on the HIV virus are recognized by helper T cell receptors and HIV binds to the surface of the host cell.
2	Host ingests viral genome and proteins, including reverse transcriptase.
3	Reverse transcriptase copies Viral RNA and makes Viral DNA.
4	Viral DNA integrates into the cellular DNA, forming a provirus. Like a prophage, the Viral DNA can be inactive for a long period of time (called latency).
5	Viral RNA is produced and host cell ribosomes translate its instructions.
6	Viruses assemble at the cell surface and new immature copies of HIV form.
7	The new HIV matures and exocytosis occurs. The virus particles are released from the cell.

Viroids and Prions

Even smaller than viruses are simple structures call viroids and prions. **Viroids** are mini-viruses (nucleotides without protein) that specifically infect plants. **Prions** are misfolded proteins that infect animals. They affect brain function and cause diseases like mad cow.

Mendel's Laws of Heredity

Genes are the basis of heredity. The German scientist Gregor Mendel first suggested the existence of genes in 1866. A gene can be pinpointed to a locus, or a particular position, on DNA. It is estimated that humans have approximately 20,000 to 25,000 genes. For any particular gene, a human inherits one copy from each parent for a total of two.

Genotype refers to the genetic makeup of an individual within a species. **Phenotype** refers to the visible characteristics and observable behavior of an individual within a species.

Genotypes are written with pairs of letters that represent alleles. **Alleles** are different versions of the same gene, and, in simple systems, each gene has one dominant allele and one recessive allele. The letter of the dominant trait is capitalized, while the letter of the recessive trait is not capitalized. An individual can be homozygous dominant, homozygous recessive, or heterozygous for a particular gene. **Homozygous** means that the individual inherits two alleles of the same type while **heterozygous** means inheriting one dominant allele and one recessive allele.

If an individual has homozygous dominant alleles or heterozygous alleles, the dominant allele is expressed. If an individual has homozygous recessive alleles, the recessive allele is expressed. For example, a species of bird develops either white or black feathers. The white feathers are the dominant allele, or trait (*A*), while the black feathers are the recessive allele (*a*). Homozygous dominant (*AA*) and heterozygous (*Aa*) birds will develop white feathers. Homozygous recessive (aa) birds will develop black feathers.

Genotype (genetic makeup)	Phenotype (observable traits)
AA	white feathers
Aa	white feathers
aa	black feathers

Chromosomal Basis of Inheritance

Tenets of Mendelian genetics:

- **The law of dominance:** Dominant alleles trump recessive alleles in phenotype (the exceptions are non-Mendelian traits)
- **The law of segregation:** Alleles for each trait are separated into gametes. One allele comes from each parent, giving the offspring two copies of each allele.
- **The law of independent assortment:** Tetrads line up in metaphase I independently of other chromosomes. Each of the 23 homologues has a 50/50 chance of being on either side.

In simple Mendelian genetics, an individual can have three different genotypes. This is shown in the table below regarding the trait of flower color:

Genotype	Referred to as	Corresponding Phenotype
PP	Homozygous dominant	Purple flowers
Pp	Heterozygous	Purple flowers
pp	Homozygous recessive	White flowers

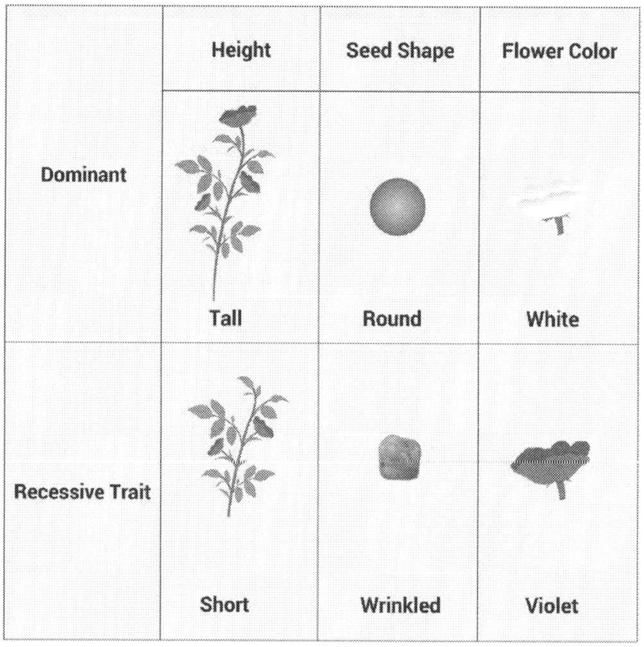

When Mendel crossed true breeds in the P generation (as shown in the Punnett square below), he noted that all the F_1 offspring had the dominant phenotype. When he crossed F1 offspring (all heterozygotes), the F2 generation consistently showed a 3:1 phenotypic ratio of purple to white flowers. These results demonstrate the first law of Mendelian genetics: the law of dominance.

Parent (P) generation: True breed cross: PP x pp

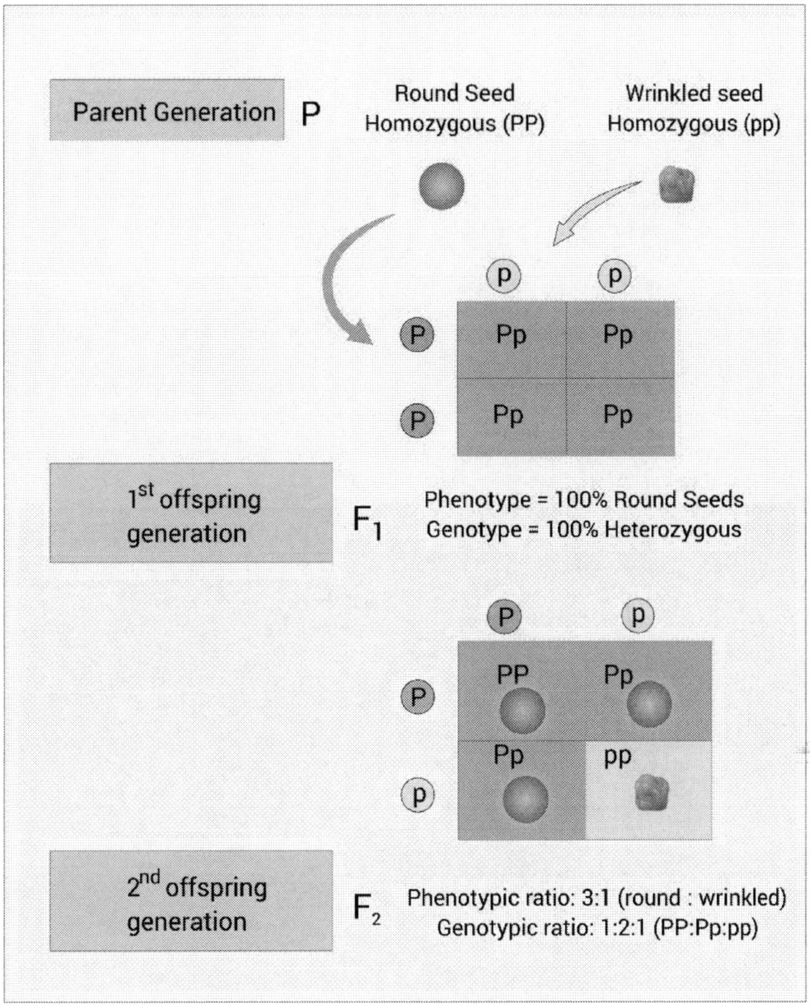

A useful application of this rule is the practice of **backcrossing**. This is when a dominant expressing organism of an unknown genotype is crossed with a recessive expressing organism. Offspring phenotype can be used to determine if the parent is homozygous or heterozygous.

Dihybrid crosses involve two traits and illustrate Mendel's law of segregation: alleles separate in meiosis.

Dihybrid crosses example: Cross PpRr x PpRr

- Purple flowers = P (dominant)
- White flowers = p (recessive)
- Round seeds = R (dominant)
- Wrinkled seeds = r (recessive)

This chart includes an example of a Dihybrid cross:

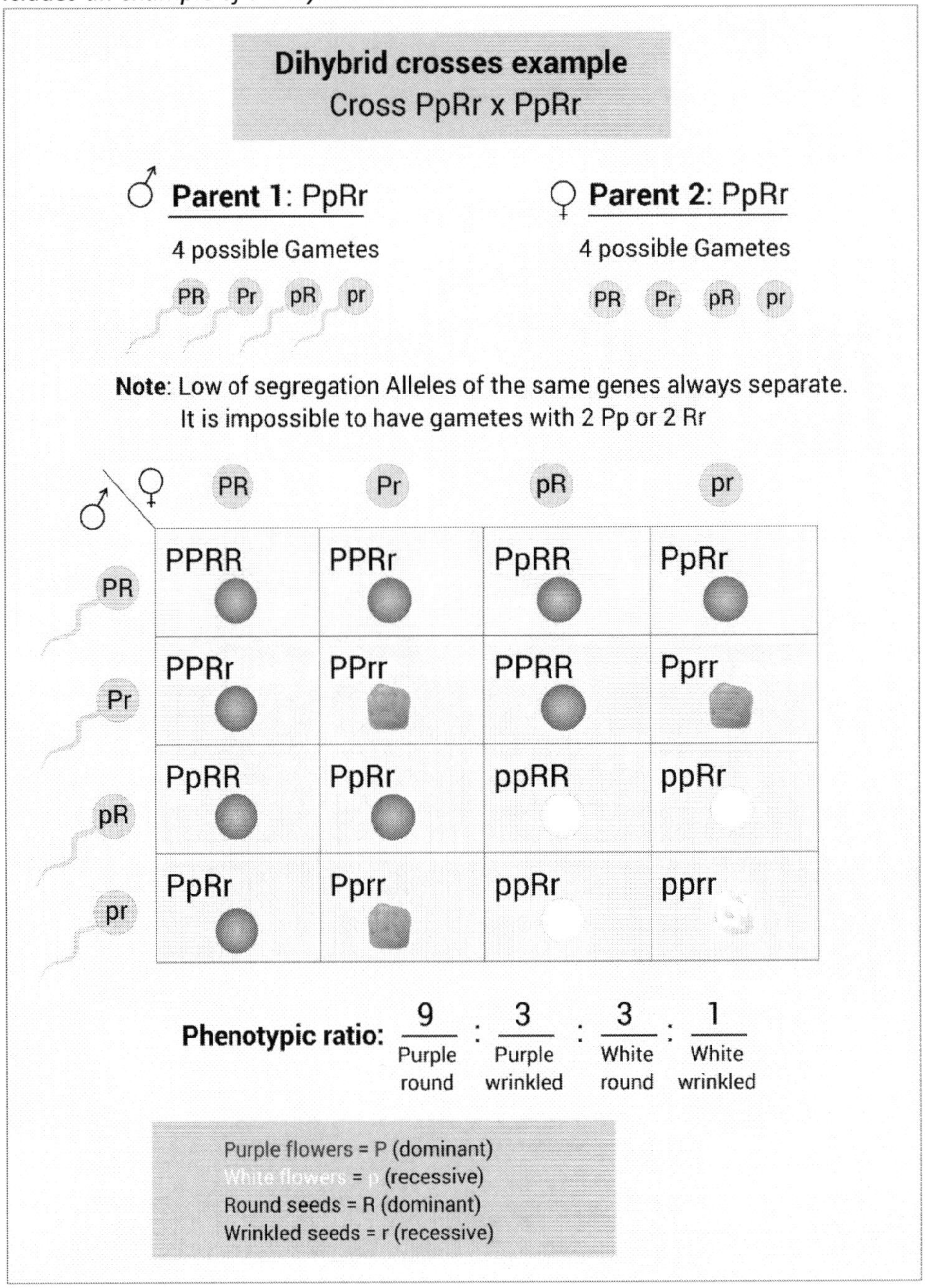

Probability can be determined by using the **law of multiplication** when all factors are present. The **law of addition** can be used to determine probability when one factor OR others may be present. Another dominant phenotype is yellow seed (Y) over green seed (y), and tall stems (T) are dominant over dwarf stems (t).

Question: What is the probability of having offspring with the genotype PPrrYytt if the parent cross is PprrYyTt x PpRryyTt?

Solution:

Pp x Pp

	P	p
P	PP	Pp
p	Pp	pp

rr x Rr

	r	r
R	Rr	Rr
r	rr	rr

Yy x yy

	Y	y
y	Yy	yy
y	Yy	yy

Tt x Tt

	T	t
T	TT	Tt
t	Tt	tt

Probabilities

$PP=\frac{1}{4}$ $rr=\frac{1}{2}$ $Yy=\frac{1}{2}$ $tt=\frac{1}{4}$

$$\frac{1}{4}\times\frac{1}{2}\times\frac{1}{2}\times\frac{1}{4}=\frac{1}{64}$$

Law of Addition (Mutually Exclusive Results)

Question: What is the probability of having offspring recessive for three different traits (could contain several different combinations)?

Solution:

	P	p
P	PP	Pp
p	Pp	pp

	r	r
R	Rr	Rr
r	rr	rr

	Y	y
y	Yy	yy
y	Yy	yy

	T	t
T	TT	Tt
t	Tt	tt

- $\frac{Pp}{pp}\,rryytt=\frac{3}{4}\times\frac{1}{2}\times\frac{1}{2}\times\frac{1}{4}=\frac{3}{64}$
- $pp\frac{RR}{Rr}\,yytt=\frac{1}{4}\times\frac{1}{2}\times\frac{1}{2}\times\frac{1}{4}=\frac{1}{64}$
- $pprr\frac{Yy}{yy}\,tt=\frac{1}{4}\times\frac{1}{2}\times\frac{1}{2}\times\frac{1}{4}=\frac{1}{64}$
- $pprryy\frac{TT}{Tt}=\frac{1}{4}\times\frac{1}{2}\times\frac{1}{2}\times\frac{3}{4}=\frac{3}{64}$

ADD • mutually exclusive

$$\frac{3}{64}+\frac{1}{64}+\frac{1}{64}+\frac{3}{64}=\frac{8}{64}\qquad\frac{8}{64}=\frac{1}{8}$$

Genetic Crosses

Genetic crosses are the possible combinations of alleles, and can be represented using Punnett squares. A **monohybrid cross** refers to a cross involving only one trait. The phenotypic ratio for a monohybrid cross is 3:1 (*AA*, *Aa*, *Aa*, *aa*) in favor of the dominant gene. This ratio occurs when both parents have a pair of dominant and recessive genes, so both parents must have heterozygous gene pairs in a monohybrid cross. If one parent has a pair of dominant genes (AA) and the other has a pair of recessive (aa) genes, the recessive trait cannot be expressed in the next generation because the resulting crosses all have the Aa genotype. A **dihybrid cross** refers to one involving more than one trait, which means more combinations are possible. The ratio of genotypes for a dihybrid cross is 9:3:3:1 when the traits are not linked, with nine offspring inheriting both dominant genes, six offspring inheriting a single dominant and a single recessive gene, and one offspring inheriting both recessive genes. The ratio for incomplete dominance is 1:2:1, which corresponds to dominant, mixed, and recessive phenotypes.

Co-Dominance and Multiple Alleles

The simple dominant/recessive model for genetics does not work for many genes.

For example, blood type is a trait that has multiple alleles: I^A, I^B, and i. I^A and I^B are "**co-dominant**" so neither is "stronger" than the other, and i is recessive to both. In the event that both co-dominant alleles are present in a genotype, both phenotypes will be present.

Genotype	Phenotype	Blood Donation Facts
IAIA, IAi	A blood (A antigens and B antibodies)	People with A blood can't receive blood from AB or B due to antibody recognition and attack of B antigen.
IAIB	AB blood (A and B antigens but no antibodies)	Universal receiver because it contains no antibodies against A or B antigens.
IBIB, IBi	B blood (B antigens and A antibodies)	People with B blood can't receive blood from AB or A due to antibody recognition and attack of A antigen.
ii	O blood (A and B antibodies)	Can only receive from other O blood (universal donor)

Blood type demonstrates the concept of co-dominance as well as multiple alleles. Below are some blood type crosses and probabilities.

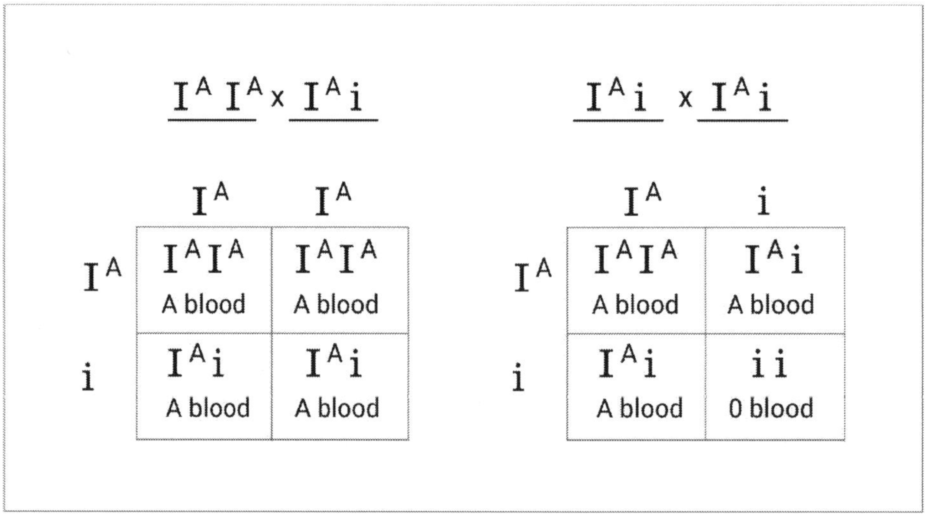

Incomplete dominance occurs when the phenotype is a blending of the two alleles instead of one being dominant over the other. For example, if black and white feathers are co-dominant in birds, heterozygous offspring will have black and white speckles. However, if black and white feathers have an incomplete dominant pattern, heterozygotes will appear grey.

Sex-Linked Genetics

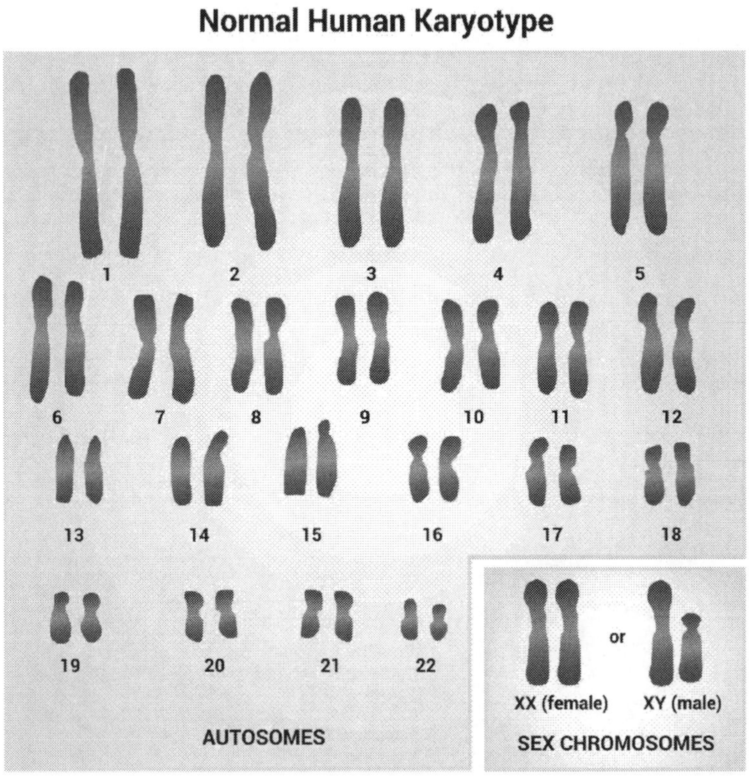

Chromosome 23 is not always homologous because it determines gender, and an individual is either XX (female) or XY (male).

The X chromosome is much larger than the Y chromosome and carries more genes, including the color-blind recessive allele. The possible genotypes, phenotypes, and inheritance patterns are shown below for the color-blind trait:

Genotype	Gender	Phenotype
XC	Female	Normal vision
XC	Female	Normal vision *She carries the allele and can pass it on to her children
Xc	Female	Color blind
XCY	Male	Normal vision
XcY	Male	Color blind

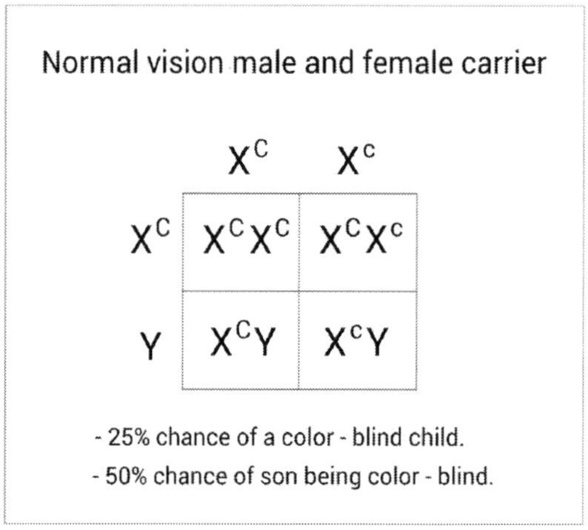

Pedigrees

Pedigrees show family ancestry and can be used to track genetic diseases through phenotypes. Circles represent females and squares represent males. Shaded shapes represent affected individuals. The genotypes of individuals can be deduced from given phenotypes. For example, below is a pedigree tracing color blindness. Individual I-1 must be X^cY because he is affected. Individual II-6 must be X^CX^c because, in order for her son to be affected, she had to pass along the color-blind allele she received from her father.

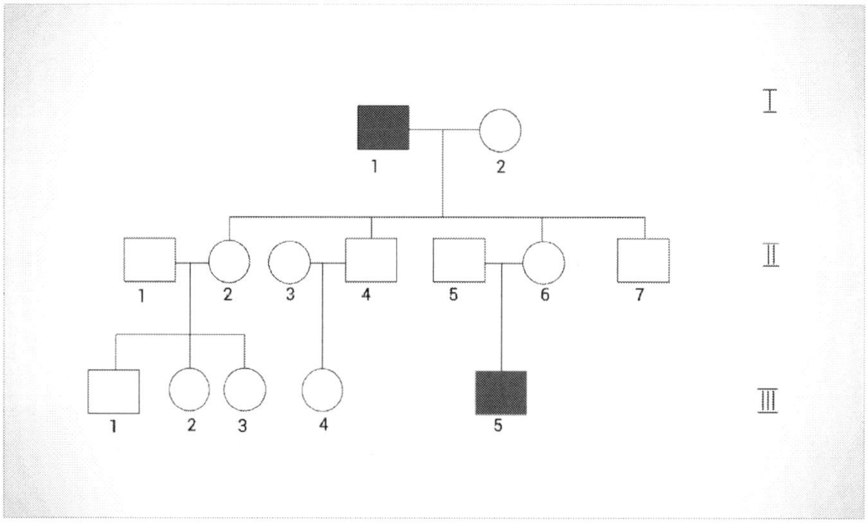

There are two types of autosomal or non-sex chromosomes (chromosomes 1-22): dominant and recessive.

The following pedigree tracing Huntington's disease is an example of an autosomal dominant disease. Every individual that has the dominant allele is affected.

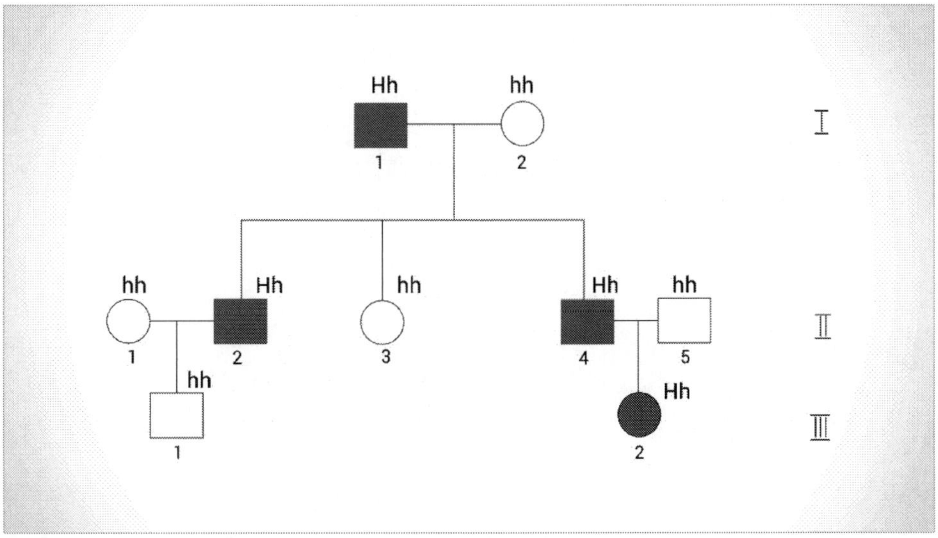

This pedigree tracing albinism is an example of an autosomal recessive disease.

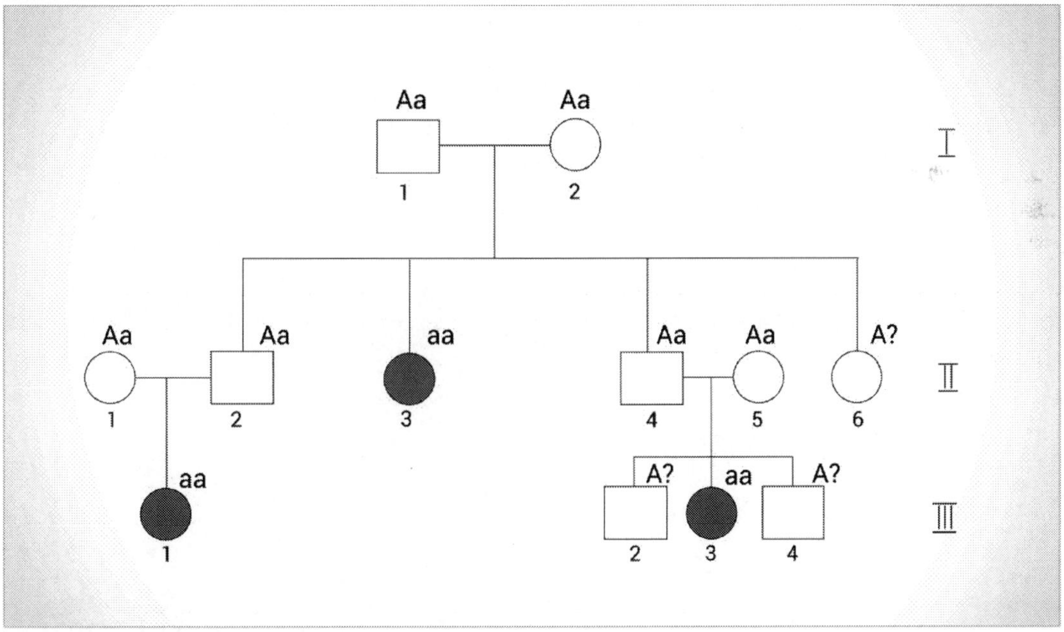

Analysis of this pedigree shows that:

- II-3, III-1, and III-3 have to be aa because they are affected.
- Any couple with children who are affected must both be heterozygotes.
- This particular pedigree can't be a sex-linked, recessive pedigree. In order to be one, an affected female must have an affected father so she can inherit his recessive allele.

Karyotypes

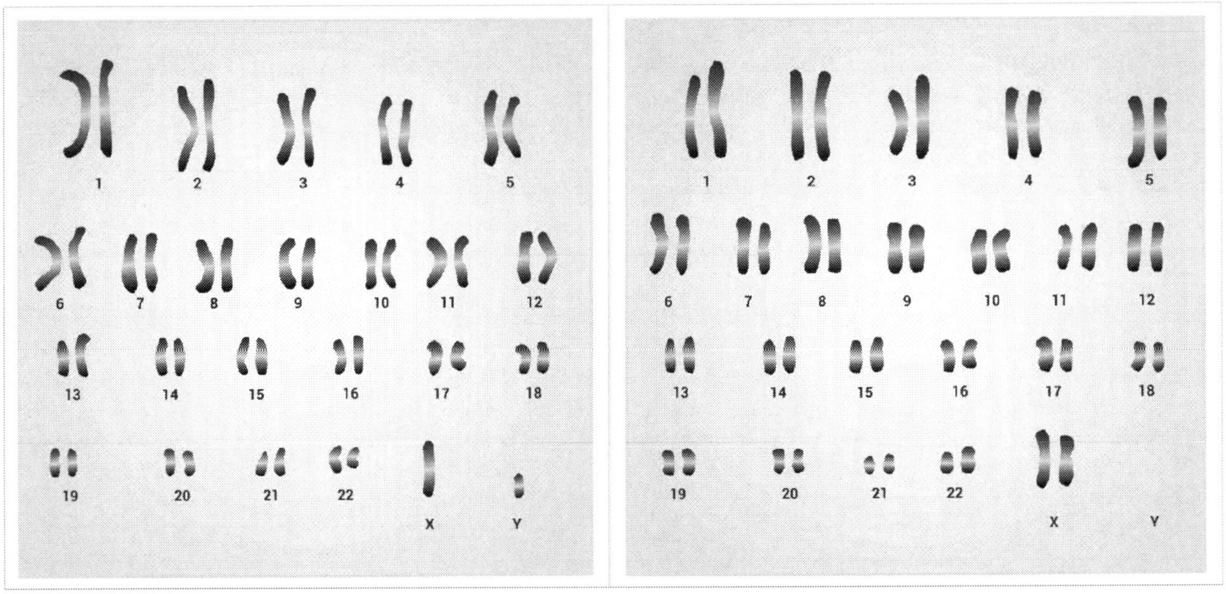

Karyotypes show a picture of an individual's 23 chromosomes and illustrate the diploid nature of our genome. Although females have two X chromosomes, only one X will be active in each cell and the other will form an inactive Barr Body. The inactive X chromosome is random. Some cells will have one inactive X chromosome while others have the second X chromosome inactive, resulting in an individual consisting of a mosaic of cells.

Karyotypes not only show gender, they can also illustrate occasional mistakes that occur in meiosis called **nondisjunction.** Nondisjunction results in improper separation of tetrads and chromatids. This results in one cell having an extra copy of a chromosome (trisomy) and another cell missing a copy of the homologous chromosome (monosomy). Fertilization with gametes affected by nondisjunction is often fatal, but some are viable.

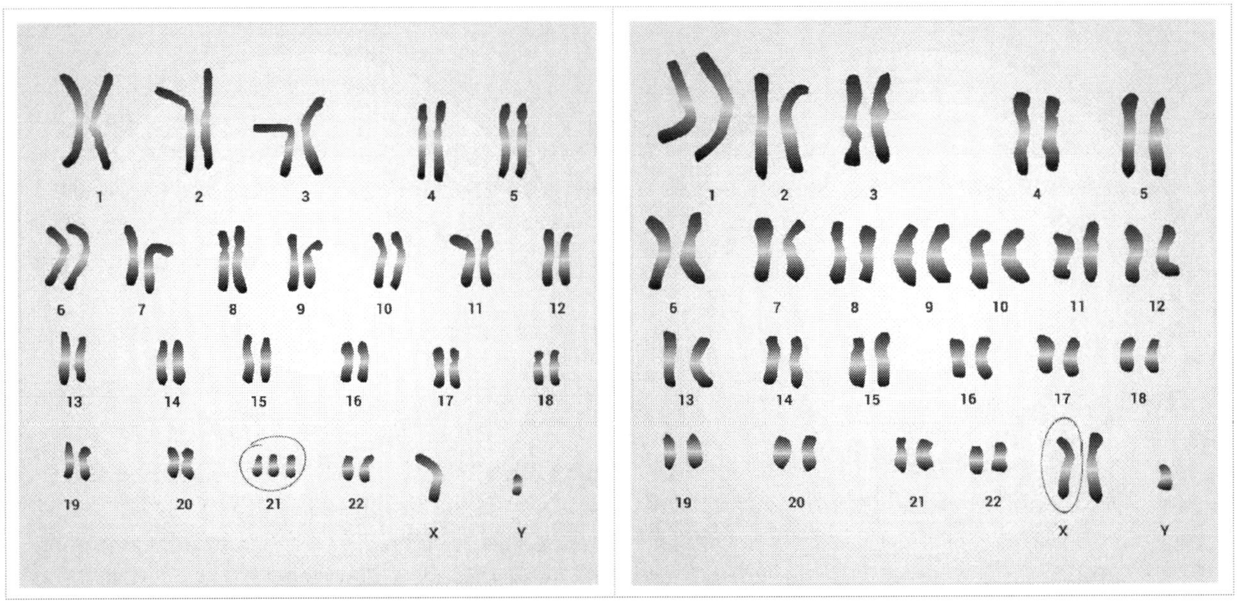

Trisomy 21: Down Syndrome **Trisomy 23: Klinefelter Syndrome**

Different diseases and their inheritance patterns are listed below:

Disease	Phenotype	Cause/Pattern
Color Blind	Red/green vision deficiency	Sex-linked recessive
Hemophilia	Blood clotting disorder	Sex-linked recessive
Muscular Dystrophy	Weak muscles and poor muscle coordination	Sex-linked recessive
Huntington's	Nervous system degeneration that has a late onset (middle age)	Autosomal dominant
Achondroplasia	Dwarfism	Autosomal dominant
Cystic Fibrosis	Excessive mucus production	Autosomal recessive
PKU	Unable to digest phenylalanine	Autosomal recessive
Tay-Sachs Disease	Intellectually disabled due to inability to metabolize lipids, death in infancy	Autosomal recessive
Sickle Cell Anemia	Red blood cell shaped like a sickle instead of a circle	Autosomal co-dominant

Disorders caused by nondisjunction and chromosomal deletions:

Disease	Phenotype	Cause/Pattern
Down Syndrome	Intellectually disabled	Trisomy 21
Klinefelter Syndrome	Sterile males	Trisomy 23 (XXY)
Turner Syndrome	Sterile females	Monosomy 21 (x)

Molecular Diversity

Diversity exists in biology, even among entities that are the same or similar, such as molecules, cells, and species. This diversity allows for a wider range of function and a greater ability to respond to the environment. One example of this is the production of antibodies by plasma cells. Antibodies are a class of proteins that provides a defense mechanism against pathogens. Their general structure is the same but each antibody has a slight variation at its tip that differentiates it for its specific pathogen. The plasma cells that make antibodies can make small changes in the antibody's DNA sequence to allow for the production of a large, diverse population of antibodies.

<u>Genetic Variation</u>

Organisms also have molecular diversity through genetic variation. One mechanism for creating genetic variation is by having two or more alternative forms, or **alleles**, of the same gene, such as with eye color or hair color. If both alleles are identical, the individual is considered **homozygous**; if the two alleles have different sequences, the individual is considered **heterozygous**. In most genes, one allele is considered more dominant than the other and will mask the appearance of the less dominant, or recessive, allele when there is a heterozygous situation.

Mutations

Another type of genetic variation that occurs is through the introduction of a **mutation**, which is a random, permanent alteration of a gene sequence. Some mutations that occur can be advantageous to an organism (and thus, the population at large should the trait get passed down), and may help the organism adapt to its environment. This may allow the organism to respond to stresses with more resilience and may even give them a fitness advantage. One example of a fitness advantage is in people whose hemoglobin gene is heterozygous for the sickle-cell trait. People who are homozygous for the sickle-cell trait have sickle-cell anemia, a harmful genetic disorder. People who are heterozygous for the sickle-cell trait are more resistant to malaria than people who are homozygous for wild-type, or normal, hemoglobin.

Gene Duplication

Genetic variation can also be established if a gene duplicates. Now the organism has an extra set of this gene, which gives the gene more chances to mutate since the original copies still maintain the necessary function of that gene. One example of this is the development of the antifreeze genes in fish that live in arctic environments, which prevents them from freezing. Originally, these fish had one copy of the sialic-acid synthase (SAS)–B gene. That gene was duplicated and, since the original SAS-B gene still retained its necessary function, the duplicated gene was no longer under selective pressure and could mutate. It then developed the mutations necessary to prevent fatal ice crystals from growing inside the fish.

Gene Expression Due to Environmental Factors

Biological variation can also exist from environmental factors influencing gene expression. Two organisms with identical genomes can have different genes turned on or off depending on their different environments. The effect of the presence of lactose on bacteria that are Lac+ is one example. Lac+ bacteria have a set of genes that metabolize the sugar lactose. When these bacteria are not around lactose, a repressor protein stops transcription of these genes. However, the presence of lactose causes the degradation of the repressor and the genes that metabolize lactose are transcribed. These environmental influences can also be seen at the organism level. Some cats have a temperature-sensitive mutation in a gene that is important for making fur color. At normal body temperature, this mutation disables the gene's function, so the fur is a light color. The cat's extremities are cooler and at a lower temperature, so the gene functions, giving the cat dark fur on its extremities.

Influence of Phenotype on Genotype

The genetic material (DNA) inherited from an individual's parents determines genotype. Natural selection leads to adaptations within a species, which affects the phenotype. Over time, individuals within a species with the most advantageous phenotypes will survive and reproduce. As result of reproduction, the subsequent generation of phenotypes receives the fittest genotype. Eventually, the individuals within a species with genetic fitness flourish and those without it are erased from the environment. This is also referred to as the concept of "survival of the fittest". When this process is duplicated over numerous generations, the outcome is offspring with a level of genetic fitness that meets or exceeds that of their parents.

Basic Atomic Structure

Structure of Matter

Elements, Compounds, and Mixtures

Everything that takes up space and has mass is composed of **matter.** Understanding the basic characteristics and properties of matter helps with classification and identification.

An **element** is a substance that cannot be chemically decomposed to a simpler substance, while still retaining the properties of the element.

Compounds are composed of two or more elements that are chemically combined. The constituent elements in the compound are in constant proportions by mass.

When a material can be separated by physicals means (such as sifting it through a colander), it is called a **mixture.** Mixtures are categorized into two types: heterogeneous and homogeneous. Heterogeneous mixtures have physically distinct parts, which retain their different properties. A mix of salt and sugar is an example of a heterogeneous mixture. With heterogenous mixtures, it is possible that different samples from the same parent mixture may have different proportions of each component in the mixture. For example, in the sugar and salt mixture, there may be uneven mixing of the two, causing one random tablespoon sample to be mostly salt, while a different tablespoon sample may be mostly sugar.

A homogeneous mixture, also called a **solution,** has uniform properties throughout a given sample. An example of a homogeneous solution is salt fully dissolved in warm water. In this case, any number of samples taken from the parent solution would be identical.

Atoms, Molecules, and Ions

The basic building blocks of matter are **atoms,** which are extremely small particles that retain their identity during chemical reactions. Atoms can be singular or grouped to form elements. Elements are composed of one type of atom with the same properties.

Molecules are a group of atoms—either the same or different types—that are chemically bonded together by attractive forces. For example, hydrogen and oxygen are both atoms but, when bonded together, form water.

Ions are electrically-charged particles that are formed from an atom or a group of atoms via the loss or gain of electrons.

Atoms that have gained or lost electrons wind up having a net electrical charge and are termed **ions**. The following are the main ions related to human health:

Bicarbonate	HCO^{3-}	A major buffer in blood. The lungs and kidneys regulate its concentration.
Chloride	Cl^{-}	Important in stomach acid and usually ingested as the salts sodium chloride (NaCl) and potassium chloride (KCl).
Calcium	Ca^{2+}	Important to muscle contraction and bone construction.
Copper	Cu	Specialized chemical reactions in the cell.
Iodine	I	Specialized chemical reactions in the cell.
Iron	Fe	Important in hemoglobin for transport of oxygen as well as part of the electron transport chain.
Magnesium	Mg^{2+}	Important in chlorophyll and animal energy production as well as a constituent of bone.
Phosphate	PO_4^{3-}	A minor intracellular pH buffer that is regulated by the kidneys and is an important factor in bone.
Potassium	K^{+}	The most plentiful mineral inside of cells and is important for nerve and muscle function.
Sodium	Na^{+}	The most common mineral outside of cells and is important for water and osmolarity regulation as well as nerve and muscle function.
Sulfate	SO_4^{2-}	A minor pH buffer for body fluids.

For the majority of light atoms, the number of protons is similar to the number of neutrons. An **isotope** is a variation of an element having the same number of protons, but a different number of neutrons. For example, all isotopes of carbon (C) have six protons. Nevertheless, C-12 has six neutrons, C-13 has seven neutrons, and C-14 has eight neutrons.

Some isotopes are radioactive and result in nuclear decay. Not all radioactive isotopes are harmful, and some are even useful to scientists and physicians. For example, C-14 is radioactive and can be used in the process of radiocarbon dating, which can be used to determine the age of organic remains. A radioactive isotope of gold (Au-198) can be utilized to treat ovarian, prostate, and brain cancer.

Basic Structure of an Atom

Atomic Models

Theories of the atomic model have developed over the centuries. The most commonly referenced model of an atom was proposed by Niels Bohr. Bohr studied the models of J.J. Thomson and Ernest Rutherford and adapted his own theories from these existing models. Bohr compared the structure of the atom to that of the Solar System, where there is a center, or nucleus, with various sized orbitals circulating around this nucleus. This is a simplified version of what scientists have discovered about atoms, including the structures and placements of any orbitals. Modern science has made further adaptations to the model, including the fact that orbitals are actually made of electron "clouds."

Atomic Structure: Nucleus, Electrons, Protons, and Neutrons

Following the Bohr model of the atom, the nucleus, or core, is made up of positively charged protons and neutrally charged neutrons. The neutrons are theorized to be in the nucleus with the protons to provide greater "balance" at the center of the atom. The nucleus of the atom makes up the majority (more than 99%) of the mass of an atom, while the orbitals surrounding the nucleus contain negatively charged electrons. The entire structure of an atom is incredibly small.

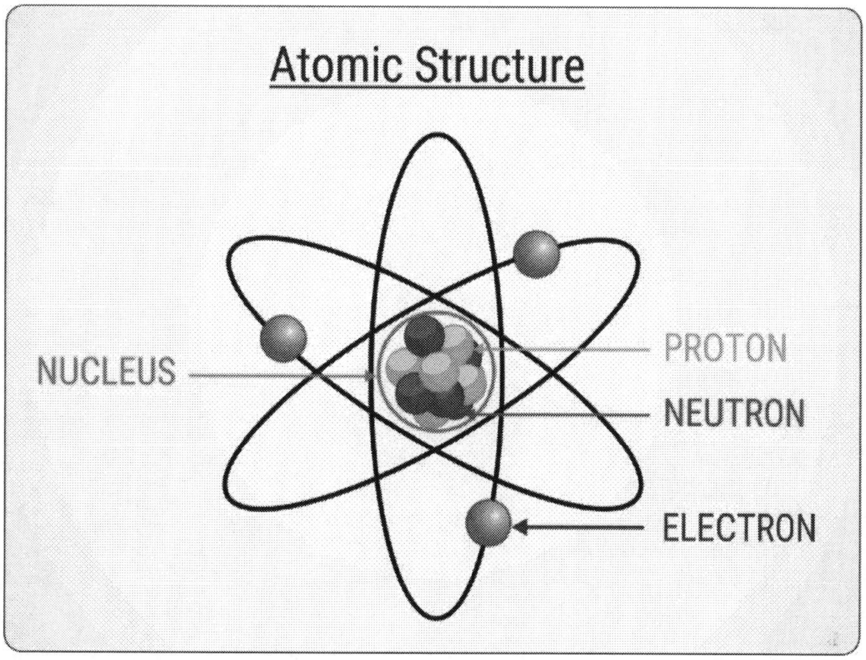

Atomic Number, Atomic Mass, and Isotopes

The **atomic number** of an atom is determined by the number of protons within the nucleus. When a substance is composed of atoms that all have the same atomic number, it is called an **element.** Elements are arranged by atomic number and grouped by properties in the periodic table.

An atom's **mass number** is determined by the sum of the total number of protons and neutrons in the atom. Most nuclei have a net neutral charge, and all atoms of one type have the same atomic number. However, there are some atoms of the same type that have a different mass number, due to an imbalance of neutrons. These are called **isotopes.** In isotopes, the atomic number, which is determined by the number of protons, is the same, but the mass number—which is determined by adding the protons and neutrons—is different due to the irregular number of neutrons.

Electron Arrangements

Electrons are most easily organized into distributions of subshells called **electron configurations.** Subshells fill from the inside (closest to the nucleus) to the outside. Therefore, once a subshell is filled, the next shell farther from the nucleus begins to fill, and so on. Atoms with electrons on the outside of a noble gas core (an atom with an electron inner shell that corresponds to the configuration of one of the noble gases, such as Neon) and pseudo-noble gas core (an atom with an electron inner shell that is similar to that of a noble gas core along with (n -1) d^{10} electrons), are called **valence electrons**. Valence electrons are primarily the electrons involved in chemical reactions. The similarities in their configurations account for similarities in properties of groups of elements. Essentially, the groups (vertical columns) on the periodic table all have similar characteristics, such as solubility and reactivity, due to their similar electron configurations.

Periodic Table of the Elements

Using the periodic table, elements are arranged by atomic number, similar characteristics, and electron configurations in a tabular format. The columns, called **groups**, are sorted by similar chemical properties and characteristics such as appearance and reactivity. This can be seen in the shiny texture of metals, the high melting points of alkali Earth metals, and the softness of post-transition metals. The rows are arranged by electron valance configurations and are called **periods.**

The elements are set in ascending order from left to right by atomic number. As mentioned, the atomic number is the number of protons contained within the nucleus of the atom. For example, the element helium has an atomic number of 2 because it has two protons in its nucleus.

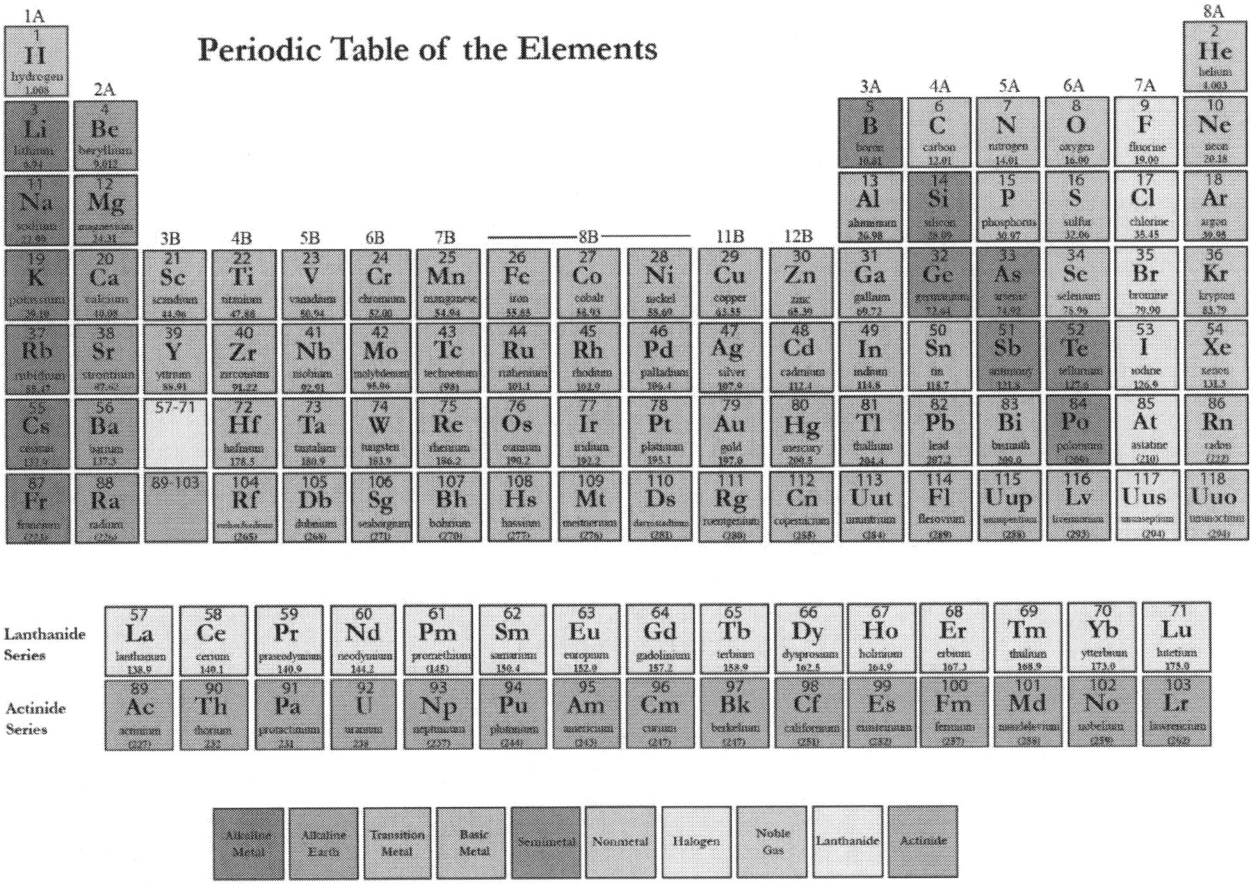

An element's mass number is calculated by adding the number of protons and neutrons of an atom together, while the atomic mass of an element is the weighted average of the naturally occurring atoms of a given element, or the relative abundance of isotopes that might be used in chemistry. For example, the atomic (mass) number of chlorine is 35; however, the atomic mass of chlorine is 35.5 amu (atomic mass unit). This discrepancy exists because there are many isotopes (meaning the nucleus could have 36 instead of 35 protons) occurring in nature. Given the prevalence of the various isotopes, the average of all of the atomic masses turns out to be 35.5 amu, which is slightly higher than chlorine's number on the periodic table. As another example, carbon has an atomic number of 12, but its atomic mass is 12.01 amu because, unlike chlorine, there are few naturally occurring isotopes to raise the average number.

Elements are arranged according to their valance electron configurations, which also contribute to trends in chemical properties. These properties help to further categorize the elements into blocks, including metals, non-metals, transition metals, alkali metals, alkali earth metals, metalloids, lanthanides, actinides, diatomics, post-transition metals, polyatomic non-metals, and noble gases. Noble gases (the far-right column) have a full outer electron valence shell. The elements in this block possess similar characteristics such as being colorless, odorless, and having low chemical reactivity. Another block, the metals, tend to be shiny, highly conductive, and easily form alloys with each other, non-metals, and noble gases.

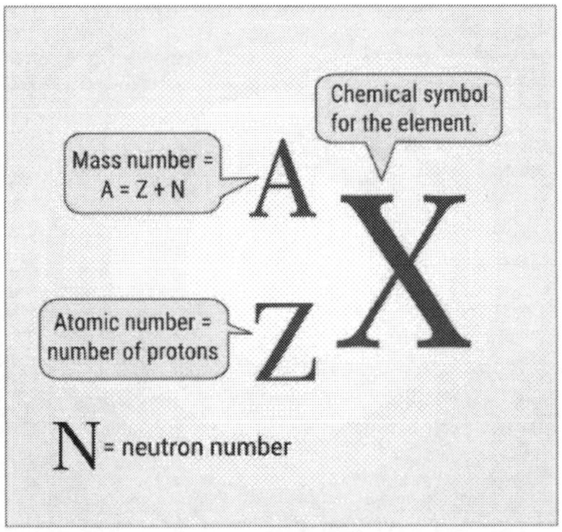

The symbols of the elements on the periodic table are a single letter or a two-letter combination that is usually derived from the element's name. Many of the elements have Latin origins for their names, and their atomic symbols do not match their modern names. For example, iron is derived from the word *ferrum*, so its symbol is Fe, even though it is now called iron. The naming of the elements began with those of natural origin and their ancient names, which included the use of the ending "ium." This naming practice has been continued for all elements that have been named since the 1940s. Now, the names of new elements must be approved by the International Union of Pure and Applied Chemistry.

The elements on the periodic table are arranged by number and grouped by trends in their physical properties and electron configurations. Certain trends are easily described by the arrangement of the periodic table, which includes the increase of the atomic radius as elements go from right to left and from top to bottom on the periodic table. Another trend on the periodic table is the increase in **ionization energy** (or the tendency of an atom to attract and form bonds with electrons). This tendency increases from left to right and from bottom to top of the periodic table—the opposite directions of the trend for the atomic radius. The elements on the right side and near the bottom of the periodic table tend to attract electrons with the intent to gain, while the elements on the left and near the top usually lose, or give up, one or more electrons in order to bond. The only exceptions to this rule are the noble gases. Since the noble gases have full valence shells, they do not have a tendency to lose or gain electrons.

Chemical reactivity is another trend identifiable by the groupings of the elements on the periodic table. The chemical reactivity of metals decreases from left to right and while going higher on the table. Conversely, non-metals increase in chemical reactivity from left to right and while going lower on the table. Again, the noble gases present an exception to these trends because they have very low chemical reactivity.

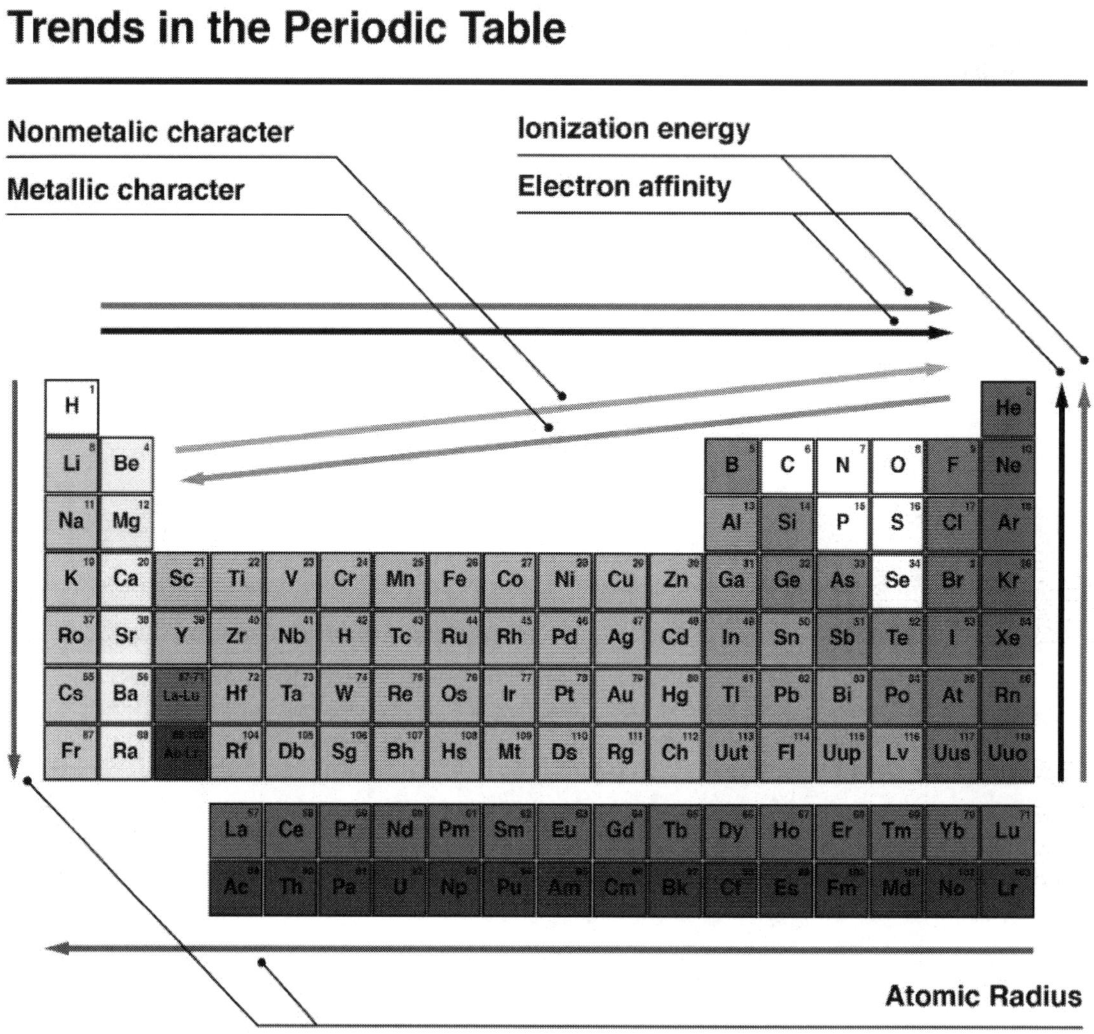

Basic Characteristics of Radioactive Materials

<u>Radioisotopes</u>
As mentioned, an isotope is a variation of an element with a different number of neutrons in the nucleus, causing the nucleus to be unstable. When an element is unstable, it will go through decay or disintegration. All manmade elements are unstable and will break down. The length of time for an unstable element to break down is called the **half-life**. As an element breaks down, it forms other elements, known as daughters. Once a stable daughter is formed, the radioactive decay stops.

<u>Characteristics of Alpha Particles, Beta Particles, and Gamma Radiation</u>
As radioactive decay is occurring, the unstable element emits alpha, beta, and gamma radiation. Alpha and beta radiation are not as far-reaching or as powerful as gamma radiation. Alpha radiation is caused by the emission of two protons and two neutrons, while beta radiation is caused by the emission of

either an electron or a positron. In contrast, gamma radiation is the release of photons of energy, not particles. This makes it the farthest-reaching and the most dangerous of these emissions.

Fission and Fusion

The splitting of an atom is referred to as fission, whereas the combination of two atoms into one is called fusion. To achieve fission and break apart an isotope, the unstable isotope is bombarded with high-speed particles. This process releases a large amount of energy and is what provides the energy in a nuclear power plant. Fusion occurs when two nuclei are merged to form a larger nucleus. The action of fusion also creates a tremendous amount of energy. To put the difference in the levels of energy between fission and fusion into perspective, the level of energy from fusion is what provides energy to the Earth's sun.

Characteristic Properties of Substances

Properties

Chemistry is the study of matter, including its properties and behavior. **Matter** is the material that the universe is made of; it is any object that occupies space and has mass. Despite the diversity of items found in the universe, matter is comprised of only about 100 substances, known as elements. Elements cannot be broken down into simpler substances. Hydrogen and oxygen are two examples of elements. When different elements join together, they form compounds. Water is a compound made from hydrogen and oxygen. Atoms and molecules are among the smallest forms of matter.

The physical and chemical properties of matter can help distinguish different substances. Physical properties include color, odor, density, and hardness. These are properties that can be observed without changing the substance's identity or composition. When a substance undergoes a physical change, its physical appearance changes but its composition remains the same. Chemical properties are those that describe the way a substance might change to form another substance. Examples of chemical properties are flammability, toxicity, and ability to oxidize. These properties are observed by changing the environment of the substance and seeing how the substance reacts. A substance's composition is changed when it undergoes a chemical change.

Many properties of matter can be measured quantitatively, meaning the measurement is associated with a number. When a property of matter is represented by a number, it is important to include the unit of measure, otherwise the number is meaningless. For example, saying a pencil measures 10 is meaningless. It could be referring to 10 of something very short or 10 of something very long. The correct measurement notation would be 10 centimeters, because a centimeter has a designated length. Other examples of properties of matter that can be measured quantitatively are mass, time, and temperature, among others.

Physical Properties vs. Chemical Properties

Both physical and chemical properties are used to sort and classify objects:

- Physical properties: refers to the appearance, mass, temperature, state, size, or color of an object or fluid; a physical change indicates a change in the appearance, mass, temperature, state, size or color of an object or fluid.
- Chemical properties: refers to the chemical makeup of an object or fluid; a chemical change refers to an alteration in the makeup of an object or fluid and forms a new solution or compound.

Differences Between Chemical and Physical Changes

A change in the physical form of matter, but not in its chemical identity, is known as a **physical change**. An example of a physical change is tearing a piece of paper in half. This changes the shape of the matter, but it is still paper.

Conversely, a **chemical change** alters the chemical composition or identity of matter. An example of a chemical change is burning a piece of paper. The heat necessary to burn the paper alters the chemical composition of the paper. This chemical change cannot be easily undone, since it has created at least one form of matter different than the original matter.

Reversible Change vs. Non-Reversible Change

Reversible change (physical change) is the changing of the size or shape of an object without altering its chemical makeup. Examples include the heating or cooling of water, change of state (solid, liquid, gas), the freezing of water into ice, or cutting a piece of wood in half.

When two or more materials are combined, it is called a mixture. Generally, a mixture can be separated out into the original components. When one type of matter is dissolved into another type of matter (a solid into a liquid or a liquid into another liquid), and cannot easily be separated back into its original components, it is called a solution.

States of matter refer to the form substances take such as solid, liquid, gas, or plasma. Solid refers to a rigid form of matter with a flexed shape and a fixed volume. Liquid refers to the fluid form of matter with no fixed shape and a fixed volume. Gas refers to an easily compressible fluid form of matter with no fixed shape that expands to fill any space available. Finally, plasma refers to an ionized gas where electrons flow freely from atom to atom.

> Examples: A rock is a solid because it has a fixed shape and volume. Water is considered to be a liquid because it has a set volume, but not a set shape; therefore, you could pour it into different containers of different shapes, as long as they were large enough to contain the existing volume of the water. Oxygen is considered to be a gas. Oxygen does not have a set volume or a set shape; therefore, it could expand or contract to fill a container or even a room. Gases in fluorescent lamps become plasma when electric current is applied to them.

Matter can change from one state to another in many ways, including through heating, cooling, or a change in pressure.

Changes of state are identified as:

- Melting: solid to liquid
- Sublimation: solid to gas
- Evaporation: liquid to gas
- Freezing: liquid to solid
- Condensation: gas to liquid
- Non-reversible change (chemical change): When one or more types of matter change and it results in the production of new materials. Examples include burning, rusting, and combining solutions. If a piece of paper is burned it cannot be turned back into its original state. It has forever been altered by a chemical change.

Properties of Water

While water is not an organic compound (meaning it does not contain carbon), it is a critical molecule for life. While it is covalent, in that the oxygen shares electrons with two hydrogen atoms, it also has polar bonds, meaning that it is slightly charged and borderline ionic. Oxygen has an atomic number of 8, and hydrogen has an atomic number of 1. Think of the oxygen nucleus as having eight positive magnets in it, and the hydrogen nucleus of having one. Now think of hydrogen having one lone electron circling around it. That electron is going to be sucked toward the more electronegative, or "powerful," oxygen nucleus, making the oxygen slightly negative. The hydrogen proton will be exposed on one side, and that part of it will be attracted to any negative oxygen neighbors. This intermolecular attraction between the hydrogen proton and a partially negative neighbor is called a hydrogen bond. Hydrogen bonds are *not* ionic; they are intermolecular attractions. The image of liquid water here shows the hydrogen bonds as dotted lines. These hydrogen bonds give water many unique properties, including the following:

- Cohesion is when water sticks to itself because of the attraction between molecules, as observed in dew.
- Adhesion is when water sticks to other structures, as observed in graduated cylinders when sides of the meniscus stick to the walls.

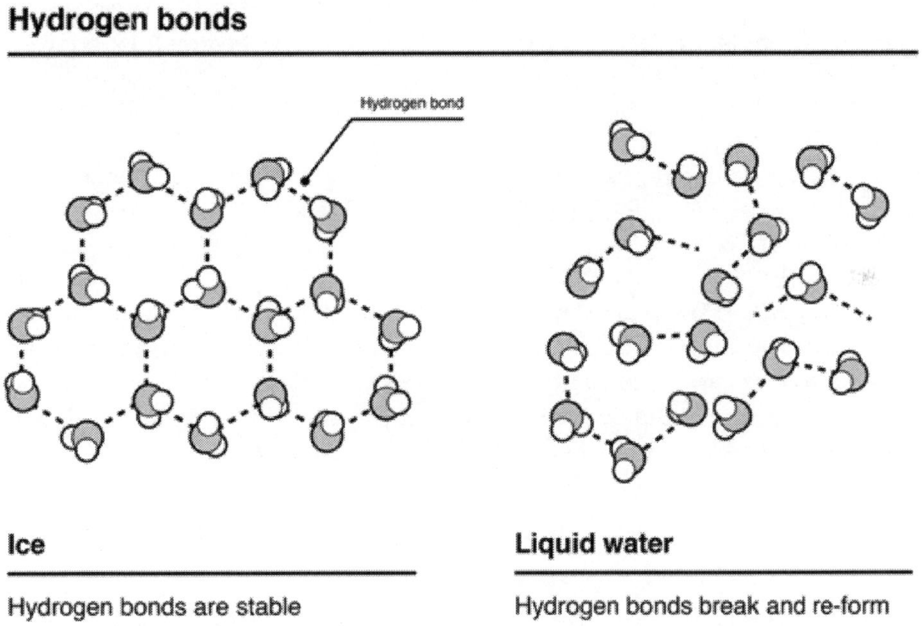

- High surface tension is when water can support some solids. Solids are supposed to be less dense than liquid, but if there is a large enough body of water, there will be a significant "sticky" film at the top due to the cohesive forces between water molecules. Heavy things can penetrate the film and sink, but lighter objects such as leaves and ice, even though they are solid, will not be able to overcome the attractive forces between the water molecules and will float.
- High boiling point is a result of the great amount of kinetic energy required to overcome water's cohesive forces and: vaporize.

- High freezing point is a result of water's attractive forces enable it to be arranged much more easily into the specific hexagonal arrangements that form ice, permitting it to solidify at a higher temperature.

- Solid ice has a very low density due to the space between the lattice/crystalline structure, which explains why ice is less dense than water, as seen in the image below.

These unique properties of water sustain life. Ice floating prevents bodies of water from freezing entirely solid because it insulates the liquid water underneath. This insulation keeps the water below it fluid and allows organisms to survive. Adhesion and cohesion of water in the xylem of plants allows water to travel great distances against the force of gravity.

Cohesion

Cohesion is the interaction of many of the same molecules. In water, cohesion occurs when there is hydrogen bonding between water molecules. Water molecules use this bonding ability to attach to each other and can work against gravity to transport dissolved nutrients to the top of a plant. A network of water-conducting cells can push water from the roots of a plant up to the leaves.

The cohesive behavior of water also causes surface tension. If a glass of water is slightly overfull, water can still stand above the rim. This is because of the unique bonding of water molecules at the surface — they bond to each other and to the molecules below them, making it seem like it is covered with an impenetrable film. A raft spider can actually walk across a small body of water due to this surface tension.

Adhesion

Adhesion is the linking of two different substances. Water molecules can form a weak hydrogen bond with, or adhere to, plant cell walls to help fight gravity.

High Specific Heat Capacity

Another important property of water is its ability to moderate temperature. Water can moderate the temperature of air by absorbing or releasing stored heat into the air. Water has the distinctive capability of being able to absorb or release large quantities of stored heat while undergoing only a small change in temperature. This is because of the relatively high specific heat of water, where specific heat is the amount of heat it takes for one gram of a material to change its temperature by 1 degree Celsius. The specific heat of water is one calorie per gram per degree Celsius, meaning that for each gram of water, it takes one calorie of heat to raise or lower the temperature of water by 1 degree Celsius.

Universal Solvent

The polarity of water molecules makes it a versatile solvent. Ionic compounds, such as salt, are made up of positively and negatively charged atoms, called cations and anions, respectively. Cations and anions are easily dissolved in water because of their individual attractions to the slight positive charge of the hydrogen atoms or the slight negative charge of the oxygen atoms in water molecules. Water molecules separate the individually charged atoms and shield them from each other so they don't bond to each other again, creating a homogenous solution of the cations and anions. Nonionic compounds, such as sugar, have polar regions, so are easily dissolved in water. For these compounds, the water molecules form hydrogen bonds with the polar regions (hydroxyl groups) to create a homogenous solution. Any substance that is attracted to water is termed hydrophilic. Substances that repel water are termed hydrophobic.

Physical and Chemical Properties of Main Group Elements and Their Compounds

Group 1, the alkali metals—apart from hydrogen (H)—are very reactive due to their single valence electron. They can react vigorously with water to form alkaline solutions and their reactivity increases from top to bottom in the group. Group 1 elements are soft and are shiny, but dull quickly after reacting with oxygen or water. They can react with halogens (Group 17), forming **ionic halides**, and with heavy chalcogens (Group 16), forming **metal chalcogenides**. They can also react with oxygen to form various compounds, such as peroxides and superoxides. All alkali metals also react with hydrogen.

Group 2, the alkaline earth metals, are similar in some ways to Group 1 metals, but they are less reactive, are harder, and have a higher melting point.

Group 13, the boron family, are metals with three electrons in their outermost shell. Boron (B) prefers an oxidation state of +3, and aluminum is the most important member of the boron family in terms of its uses. Aluminum also usually adopts a +3-oxidation state and is a soft, malleable metal that is silver in color. Aluminum does not react with water due to an **anodizing effect**—a protective layer of Al_2O_3, which protects against oxidization. Gallium (Ga) and indium (In) also prefer an oxidation state of 3+, whereas thallium (Tl) most commonly has an oxidation state of +1.

Group 14 is the carbon family. Each element in this group contains only two electrons in its outer *p* orbital. The Group 14 elements tend to adopt oxidation states of +4, or, for the heavier compounds, +2. Carbon can, and often does, form double and triple bonds with itself or other elements, but the heavier elements in this group can only form single bonds. Metallic properties increase down the group. Carbon (C) is not a metal. Silicon (Si) and germanium (Ge) are metalloids. Tin (Sn) and lead (Pb) are metals. Carbon exists in three allotropes—graphite, diamond, and fullerenes—each of which has different physical properties. Graphite has lubricating properties. Diamond is hard and can dissipate heat well, and **fullerenes**—carbon rings joined together into more complex structures like buckeyballs—can be used to create carbon nanotubes.

Group 15 is the nitrogen family. All Group 15 elements have the outer electron shell configuration of ns^2np^3. Group 15 elements generally follow the periodic trends: decreasing electronegativity going down the group, decreasing ionization energy, increasing atomic radii, decreasing electron affinity, increasing melting and boiling points, and increasing metallic character. Nitrogen (N) and phosphorous (P) are nonmetals. Arsenic (As) is a highly poisonous metalloid, and antimony (Sb) is a metalloid. Bismuth (Bi) is metallic. Group 15 elements exist in various different oxidation states.

Group 16 is the oxygen family—also called the **chalcogen family**. These elements have two electrons in the outermost *s* orbital and four in the outermost *p* orbital. Periodic table trends are generally followed in this group, and the most common oxidation state is -2, although +2, +4, and +6 states are also possible for some members. At room temperature, oxygen (O) is a gas, sulfur (S) is a yellow solid, selenium (Se) is a red/black amorphous solid or red/grey crystal, tellurium (Te) is a silver / white metalloid, and polonium (Po) is a rare, highly toxic, radioactive compound.

Group 17 is the halogens. They are the most reactive nonmetals due to their electron configuration p^5—almost full, which is very reactive. They react with most metals to produce **salts** (ionic compounds), such as copper chloride and sodium chloride. Halogens exist in all states of matter. At room temperature, iodine (I) and asinine (At) are solids, bromine (Br) is a liquid, and fluorine (F) and chlorine (Cl) are gases.

Group 18 is the noble gases. They have a full octet in their outer energy level, making them very stable. For example, neon (Ne) has an electron configuration of $2s^2 2p^6$. All end in p^6, which is a very stable electron configuration. Helium and neon, in particular, have very low reactivity.

Changes in States of Matter

States of Matter

Matter exists in phases and is mostly in the solid, liquid, or gas form. The **Kinetic Theory of Matter** states that matter is composed of a large number of small particles (specifically, atoms and molecules) that are in constant motion. The distance between the separations in these particles determines the state of the matter: solid, liquid, or gas. The constant movement of particles causes them to bump into each other, thus allowing the particles to transfer energy between each other. This bumping and transferring of energy helps explain the transfer of heat and the relationship between pressure, volume, and temperature.

Solid matter has particles packed closely together. Solids are not compressible because they are already so compact that they are actually vibrating due to the significant attractive and repulsive intermolecular forces. Very strong intermolecular forces contribute to a solid having a definite shape and volume. Solids have no flow and do not take the shape of their container.

Liquids, on the other hand, have flow because particles are not as tightly packed. They take the shape of their container but are not easily compressible because the particles are still fairly close together. The molecules in a liquid are not in an orderly arrangement and can move past one another. Weak intermolecular forces contribute to a liquid having an indefinite shape, but definite volume.

The molecules in a gas have a large amount of space between them. A gas will diffuse indefinitely if unconfined, while it will assume the shape and volume of its container if enclosed. In other words, a gas has no definite shape or volume. Gas particles are different from solids and liquids in that they are very easily compressible. They are so far apart that intermolecular forces are negligible. Lastly, a gas is flows quite easily.

The Ideal Gas Law states that pressure, volume, and temperature are all related through the equation: $PV = nRT$, where P is pressure, V is volume, n is the amount of the substance in moles, R is the gas constant, and T is temperature.

Through this relationship, volume and pressure are both proportional to temperature, but pressure is inversely proportional to volume. Therefore, if the equation is balanced, and the volume decreases in the system, pressure needs to proportionately increase to keep both sides of the equation balanced. In contrast, if the equation is unbalanced and the pressure increases, then the temperature would also increase, since pressure and temperature are directly proportional.

Phase Transitions

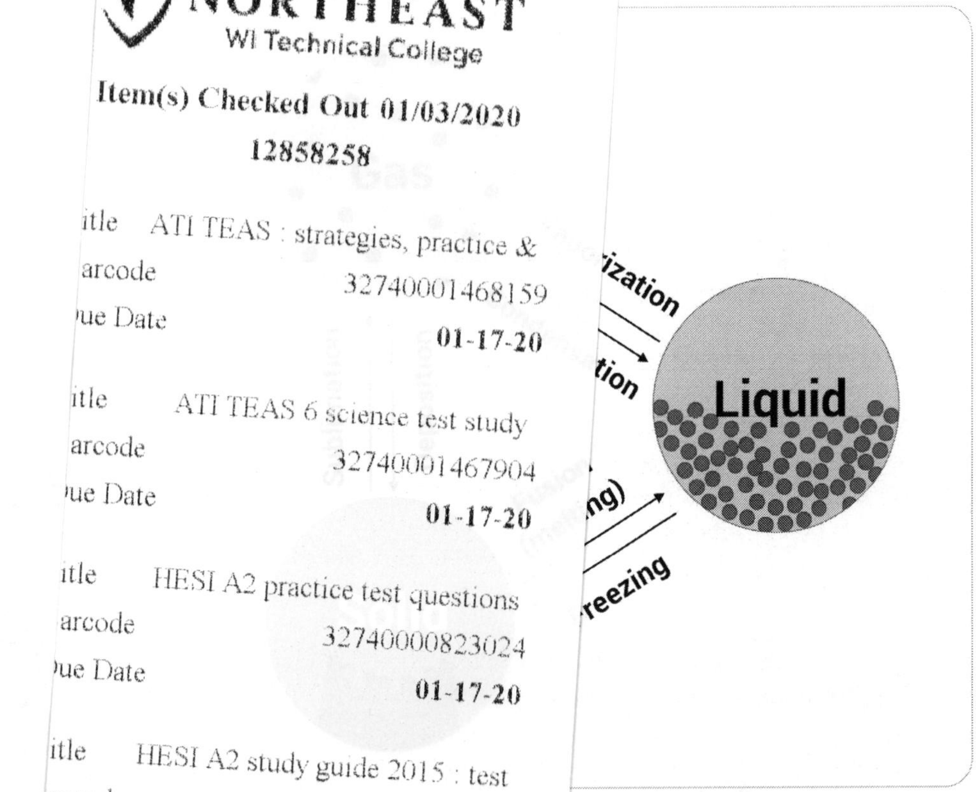

Phase changes are best described in terms of temperature and kinetic energy. Consider a tightly-packed solid that has particles very close together. If heat is added, kinetic energy of the molecules increase. If enough heat is added, the particles will move quickly enough that they will begin to slide past each other and form a liquid. Moving from a solid to a liquid is called melting. The opposite change is freezing. Freezing occurs when liquid particles move slower and pack closely together into a solid.

Think of liquid water heating on a stove. When the boiling point is reached, some of the particles will turn [illegible] energy. Moving from this liquid from into a gas is called **vaporization**. There are two types of vaporization—evaporation and boiling. **Evaporation** is a surface phenomenon and involves the conversion of liquid into a gas below the boiling temperature at a given pressure. Evaporation is also an important component of the water cycle. **Boiling** occurs below the surface and involves the conversion of liquid into a gas at or above the boiling temperature. The opposite reaction is **condensation**, which occurs when gas particles slow down and form a liquid. This is seen on the exterior of a cold drink when water vapor in the air cools on the glass and forms water beads. It is the reverse of evaporation. Condensation is also most often synonymous with the water cycle. [illegible]

Less common are direct changes from solid to gas, or **sublimation**, or from a gas to a solid, or **deposition**. Sublimation appears as a solid emitting a cloud, like dry ice. Deposition can happen with water in extremely cold temperatures. Frost is water that has gone directly from a gas to a solid.

The amount of energy needed to change matter from one state to another is labeled by the terms for phase changes. For example, the temperature needed to supply enough energy for matter to change from a liquid to a gas is called the **heat of vaporization**. When heat is added to matter in order to cause a change in state, there will be an increase in temperature until the matter is about to change its state. During its transition, all of the added heat is used by the matter to change its state, so there is no increase in temperature. Once the transition is complete, then the added heat will again yield an increase in temperature.

Each state of matter is considered to be a phase, and changes between phases are represented by phase diagrams. These diagrams show the effects of changes in pressure and temperature on matter. The states of matter fall into areas on these charts called **heating curves**.

Chemical Reactions

Types of Chemical Reactions

Chemical reactions are characterized by a chemical change in which the starting substances, or reactants, differ from the substances formed, or products. Chemical reactions may involve a change in color, the production of gas, the formation of a precipitate, or changes in heat content. The following are the five basic types of chemical reactions:

Reaction Type	**Definition**	**Example**
Decomposition	A compound is broken down into two or more smaller elements or compounds.	$2H_2O \rightarrow 2H_2+O_2$
Synthesis	Two or more elements or compounds are joined together.	$2H_2+O_2 \rightarrow 2H_2O$
Single Displacement	A single element or ion takes the place of another in a compound. Also known as a substitution reaction.	$Zn+2HCl \rightarrow ZnCl_2+H_2$
Double Displacement	Two elements or ions exchange a single atom each to form two different compounds, resulting in different combinations of cations and anions in the final compounds. Also known as a metathesis reaction.	$H_2SO_4+2NaOH \rightarrow Na_2So_4+2H_2O$
Oxidation-Reduction	Elements undergo a change in oxidation number. Also known as a redox reaction.	$2S_2O_3^{2-}+I_2 \rightarrow S_4O_6^{2-}+2I^-$
Acid-Base	Involves a reaction between an acid and a base, which usually produces a salt and water	$HBr+NaOH \rightarrow NaBr+H_2O$
Combustion	A hydrocarbon (a compound composed of only hydrogen and carbon) reacts with oxygen to form carbon dioxide and water.	$CH_4+2O_2 \rightarrow CO_2+2H_2O$

Balancing Chemical Reactions

Chemical reactions are expressed using chemical equations. Chemical equations must be balanced with equivalent numbers of atoms for each type of element on each side of the equation. Antoine Lavoisier, a French chemist, was the first to propose the Law of Conservation of Mass for the purpose of balancing a chemical equation. The law states, "Matter is neither created nor destroyed during a chemical reaction.":

The reactants are located on the left side of the arrow, while the products are located on the right side of the arrow. Coefficients are the numbers in front of the chemical formulas. Subscripts are the numbers to the lower right of chemical symbols in a formula. To tally atoms, one should multiply the formula's coefficient by the subscript of each chemical symbol. For example, the chemical equation $2H_2 + O_2 \rightarrow 2H_2O$ is balanced. For H, the coefficient of 2 multiplied by the subscript 2 = 4 hydrogen atoms. For O, the coefficient of 1 multiplied by the subscript 2 = 2 oxygen atoms. Coefficients and subscripts of 1 are understood and never written. When known, the form of the substance is noted with (g)=gas, (s)=solid, (l)=liquid, or (aq)=aqueous.

Catalysts

Catalysts are substances that accelerate the speed of a chemical reaction. A catalyst remains unchanged throughout the course of a chemical reaction. In most cases, only small amounts of a catalyst are needed. Catalysts increase the rate of a chemical reaction by providing an alternate path requiring less activation energy. **Activation energy** refers to the amount of energy required for the initiation of a chemical reaction.

Catalysts can be homogeneous or heterogeneous. Catalysts in the same phase of matter as its reactants are homogeneous, while catalysts in a different phase than reactants are heterogeneous. It is important to remember catalysts are selective. They don't accelerate the speed of all chemical reactions, but catalysts do accelerate specific chemical reactions.

Enzymes

Enzymes are a class of catalysts instrumental in biochemical reactions, and in most, if not all, instances are proteins. Like all catalysts, enzymes increase the rate of a chemical reaction by providing an alternate path requiring less activation energy. Enzymes catalyze thousands of chemical reactions in the human body. Enzymes are proteins and possess an active site, which is the part of the molecule that binds the reacting molecule, or substrate.

The "lock and key" analogy is used to describe the substrate key fitting precisely into the active site of the enzyme lock to form an enzyme-substrate complex.

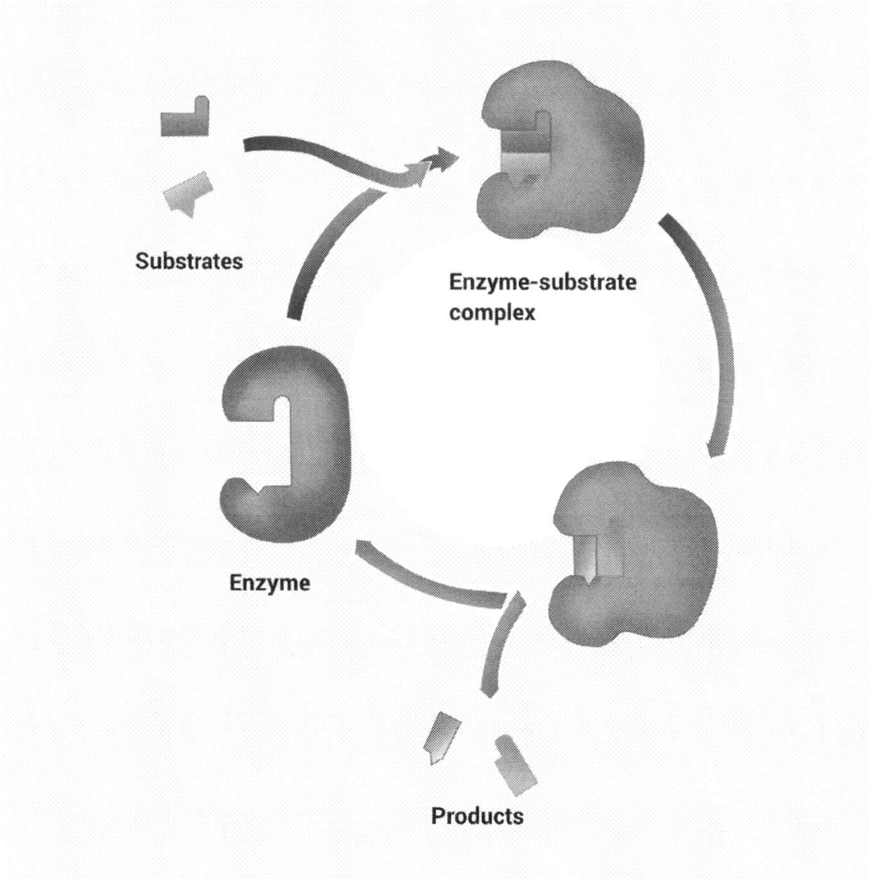

Many enzymes work in tandem with cofactors or coenzymes to catalyze chemical reactions. **Cofactors** can be either inorganic (not containing carbon) or organic (containing carbon). Organic cofactors can be either coenzymes or prosthetic groups tightly bound to an enzyme. **Coenzymes** transport chemical groups from one enzyme to another. Within a cell, coenzymes are continuously regenerating and their concentrations are held at a steady state.

Several factors including temperature, pH, and concentrations of the enzyme and substrate can affect the catalytic activity of an enzyme. For humans, the optimal temperature for peak enzyme activity is approximately body temperature at 98.6 ^{0}F, while the optimal pH for peak enzyme activity is approximately 7 to 8. Increasing the concentrations of either the enzyme or substrate will also increase the rate of reaction, up to a certain point.

The activity of enzymes can be regulated. One common type of enzyme regulation is termed **feedback inhibition,** which involves the product of the pathway inhibiting the catalytic activity of the enzyme involved in its manufacture.

Using available free energy, reactions occur within an organism that create organization and decrease entropy.

The below graphs illustrate how free energy is used in two basic biochemical reactions:

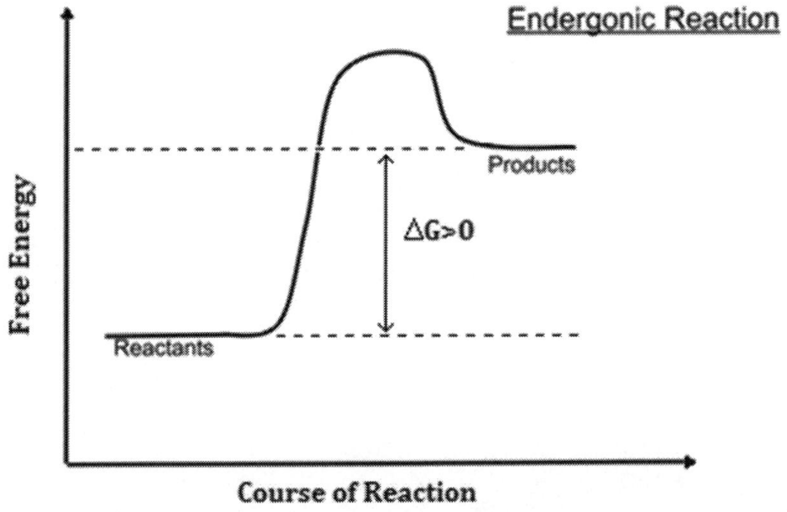

The image above illustrates a reaction that decreases entropy (or increases order), called an **endergonic reaction**. Endergonic, or anabolic, reactions occur when reactants absorb energy from the surroundings so that the products hold more energy than the reactants. Anabolic reactions enable the organism to make bigger things (polymers) from smaller things (monomers), such as forming new cells, or making proteins from amino acids.

Conversely, the image below illustrates an exergonic, or catabolic, reaction that releases energy. In this reaction, the products hold less energy than the reactants, and entropy (or disorder) is increased. Catabolic reactions enable organisms to break down bigger things (polymers) into smaller things (monomers); for example, breaking down proteins into their respective amino acids.

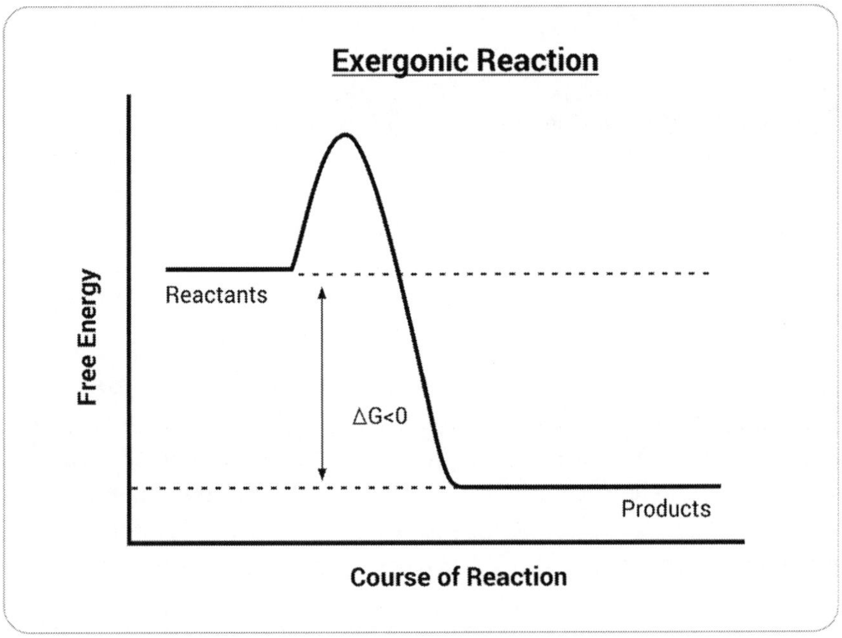

pH, Acids, and Bases

pH refers to the power or potential of hydrogen atoms and is used as a scale for a substance's acidity. In chemistry, pH represents the hydrogen ion concentration (written as $[H^+]$) in an aqueous, or watery, solution. The hydrogen ion concentration, $[H^+]$, is measured in moles of H^+ per liter of solution.

The pH scale is a logarithmic scale used to quantify how acidic or basic a substance is. pH is the negative logarithm of the hydrogen ion concentration: $pH = -\log [H^+]$. A one-unit change in pH correlates with a ten-fold change in hydrogen ion concentration. The pH scale typically ranges from zero to 14, although it is possible to have pHs outside of this range. Pure water has a pH of 7, which is considered neutral. pH values less than 7 are considered acidic, while pH values greater than 7 are considered basic, or alkaline.

Below is an image of the pH scale:

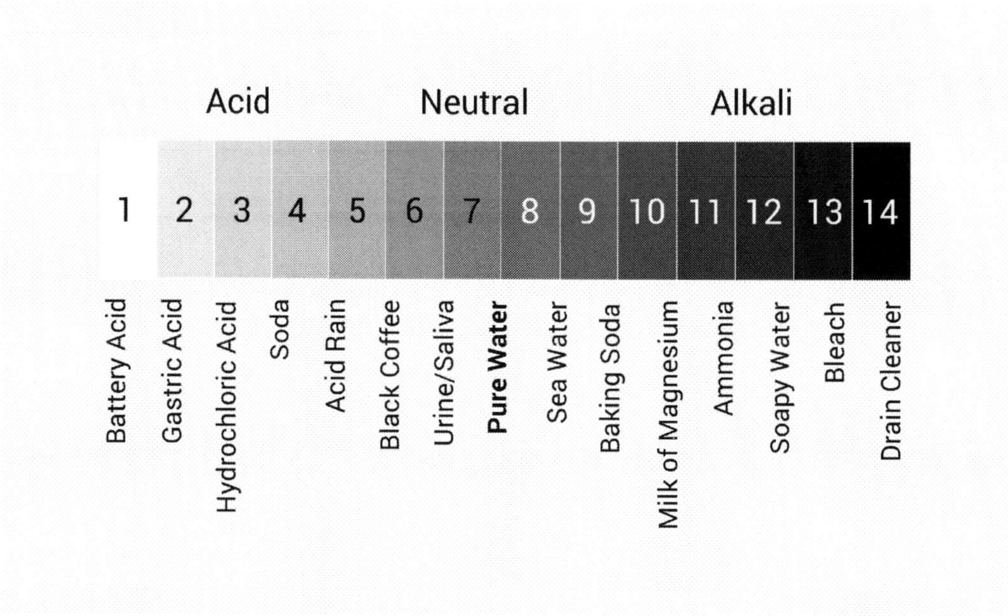

Generally speaking, an acid is a substance capable of donating hydrogen ions, while a base is a substance capable of accepting hydrogen ions. A buffer is a molecule that can act as either a hydrogen ion donor or acceptor. Buffers are crucial in the blood and body fluids, and prevent the body's pH from fluctuating into dangerous territory. pH can be measured using a pH meter, test paper, or indicator sticks.

Electrons and Chemical Bonds

Electrons orbit the nucleus in atomic shells, or electron clouds. For example, the first atomic shell can accommodate two electrons, the second atomic shell can hold a maximum of eight electrons, and the third atomic shell can house a maximum of eight electrons. The negatively charged electrons orbiting the nucleus are attracted to the positively charged protons in the nucleus via electromagnetic force. The attraction of opposite electrical charges gives rise to chemical bonds, which refers to the ways atoms are attached to each other.

Chemical bonding typically results in the formation of a new substance, called a compound. Only the electrons in the outermost atomic shell are able to form chemical bonds. These electrons are known as valence electrons, and they are what determines the chemical properties of an atom.

Chemical bonds refer to the manner in which atoms are attached to one another. Atoms may be held together with three fundamental types of chemical bonds—ionic, covalent, or hydrogen.

Ionic Bonding

In an ionic bond, an atom loses one or more electrons to another who gains them. The atoms do this so that they can achieve a full outermost shell of electrons, which is the configuration with the lowest amount of energy, and these are typically the strongest types of bonds. Ionic bonds typically occur in metals and nonmetals to form salts like $MgCl_2$ and BF_3. When an atom has a different number of electrons than its number of protons, it is termed an ion. A positively charged ion is referred to as a **cation**, while a negatively charged ion is referred to as an **anion.** For example, a sodium cation (Na^{+1}) can form an ionic bond to a chlorine anion (Cl^{-1}) to form sodium chloride (NaCl), which is also known as table salt. The majority of the molecules formed with ionic bonds such as this have a crystalline structure.

Covalent Bonding

In a covalent bond, two or more atoms share one or more electrons. Covalent bonds are the most plentiful type of bond making up the human body. They are typically found in molecules containing carbon. Only six elements typically form covalent bonds: carbon (C), nitrogen (N), phosphorus (P), oxygen (O), sulfur (S), and hydrogen (H). Covalent bonds are a crucial source of energy for living organisms. Covalent bonds are comparable in strength to ionic bonds, but stronger than hydrogen bonds, and are typically used to bind the basic macromolecules—carbohydrates, lipids, nucleic acids, and proteins—together.

Hydrogen Bonding

Hydrogen bonds are temporary and weak. They typically occur between two partial, opposite electrical charges. For example, hydrogen bonds form when a hydrogen (H) atom in the vicinity of nitrogen (N), fluorine (F), or oxygen (O) atoms. These partial electrical charges are called **dipoles,** and are caused by the unequal sharing of electrons between covalent bonds. Water is the most prevalent molecule that forms hydrogen bonds.

Hydrogen bonds contribute to the adhesiveness and cohesiveness properties of molecules like water. Adhesiveness confers glue-like properties to molecules which ensure they stick or connect more easily with other molecules, similar to wetting a suction cup before sticking it to a surface. Cohesiveness refers to a molecule's ability to form hydrogen bonds with itself. For example, the cohesiveness of water is the reason why it has a high boiling point, which is a physical property.

Energy in Biological Systems

As long as the energy released (exergonic) exceeds the energy absorbed (endergonic) in an organism, it will continue to function and sustain life. If there is not energy available from exergonic reactions to drive endergonic reactions, the organism will die, since energy is required to perform all metabolic functions. **Metabolism** is the set of processes carried out by an organism that permits the exchange of energy between itself and the environment, enabling it to change and grow internally.

In other words: metabolism = catabolism + anabolism

- If catabolism is greater than anabolism, i.e., energy released is greater than energy consumed, then an organism can live.
- If catabolism is less than anabolism, i.e., energy released is less than energy consumed, then an organism will die.

Organisms must consume energy, in the form of food or light, to perform anabolic reactions. If they are autotrophs, such as plants, they produce their own food. If they are heterotrophs, such as animals, they absorb the energy provided by food, most commonly sugars. The most common form of sugar used for energy is the simple sugar glucose, $C_6H_{12}O_6$, which contains extremely high amounts of potential energy stored within its atomic bonds.

Exergonic Reactions Can Drive Endergonic Reactions

Cells balance their energy resources by using the energy from exergonic reactions to drive endergonic reactions forward, a process called **energy coupling**. Adenosine triphosphate, or ATP, is a molecule that is an immediate source of energy for cellular work. When it is broken down, it releases energy used in endergonic reactions and anabolic pathways. ATP breaks down into adenosine diphosphate, or ADP, and a separate phosphate group, releasing energy in an exergonic reaction. As ATP is used up by reactions, it is also regenerated by having a new phosphate group added onto the ADP products within the cell in an endergonic reaction.

Even within a single process, there are energy investments and divestments in many necessary reactions along the way. ATP is necessary for all of these reactions. It can be made in two different ways: substrate-level phosphorylation and oxidative phosphorylation.

Substrate-level phosphorylation occurs when a phosphoryl group (PO_3^{2-}) is donated to an ADP (adenosine diphosphate) molecule from a substrate by using enzymes, thereby creating ATP. In oxidative phosphorylation, a free phosphate joins ADP via the energy provided by chemiosmosis and the mechanical movement of the ATP synthase (the enzyme that creates ATP).

Obtaining energy at the cellular level and understanding how energy investment is used to create organization in living things, is important in order to understand energy exchange between all of the cells of an organism and even between an organism and its environment.

Scientific Reasoning

Scientific Measurement Using Laboratory Tools

Technology and mathematics are crucial to the advancement of scientific investigations and communications. Individuals in specialized scientific fields need answers to problems swiftly, as they are frequently under time constraints. Accurate and prompt scientific investigation is dependent on technological and mathematical skills and their advancement. Advances in technology and mathematics can help accelerate the pace of scientific investigations and communications.

Although scientific investigations can be communicated to some degree through subjective qualitative data, objective quantitative data is preferred to establish rigorous mathematical conclusions. Connections can also be conveyed through the use of graphs, formulas, and equations. Mathematical relationships can be proven with numerical models.

Software like spreadsheets and networks like the Internet have expanded scientific communication by providing multiple avenues to distribute information. Increased computing capacity and fiber optics have also increased the speed and processing time of scientific communication.

Technology and Mathematics in Science Research

There are many different reasons for including technology and mathematics in science research. Technology is always advancing and its application is playing an increasingly larger role in science research. Technology enhances the gathering and manipulation of scientific data. Technology also increases the accuracy and precision of scientific data, which in many ways is becoming more dependent on the type of technology adopted. Graphs or models can be produced to highlight emerging patterns or trends in scientific data which can be used to improve the interpretation and understanding of specific scientific data from the past or present.

Mathematics has been around for centuries and has constantly been called upon in scientific research to help understand and explain the workings of the natural world. It is a crucial part of the scientific method. Mathematics can help improve and refine the asking of questions or hypotheses in a scientific argument. Advances in mathematics have led to improvements in gathering data, constructing scientific explanations, and communicating results of scientific investigations. Lastly, mathematics has improved the efficiency of processing and manipulating often complicated scientific data.

When recording data, it is of the utmost importance to also record units. Acceptable customary units in the U.S. include feet, gallon, and pound which measure length, volume, and mass, respectively. Laboratory data is typically recorded with SI units which are universally used. Standard SI units include the following:

- Second (s) for time
- Meter (m) for length
- Gram (g) for mass
- Liter (L) for volume
- Celsius (oC): – or – Kelvin (K) for temperature (oC + 273 = K)

Prefixes also indicate the scale, as seen in the chart below.

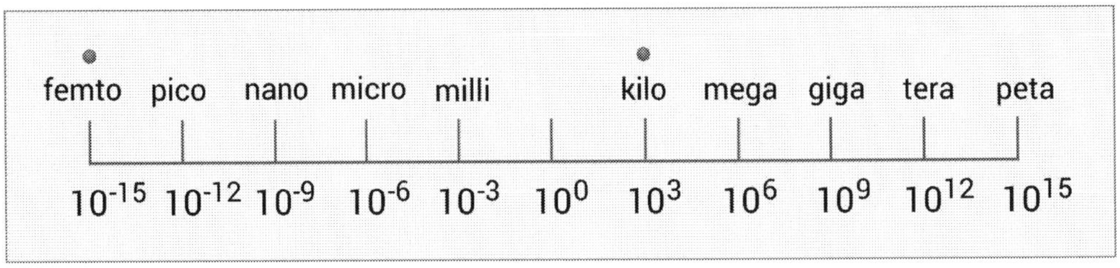

Other important concepts regarding collecting data are scientific notation, rounding, and significant figures.

Scientific notation should be used when appropriate. The correct form is to have the ones unit followed by the decimal point with the appropriate amount of significant figures followed by an indication of the power of ten.

1,320,000 = 1.32×10^6

0.000867 = 8.67×10^{-4}

Rounding should be used to the correct number of significant figures. 0-4 always keeps the preceding number as is while 5-9 raises it.

1.325 rounds to 1.33
4.289 rounds to 1.29
5.998 rounds to 6.00

Significant figures should be used, meaning that measurements too precise to be recorded should not be included. This concept is illustrated in the following scenarios.

A balance that goes to the hundredths place measures a rock as 1.92. It would be incorrect to record the value as 2 or 1.9 (remember rounding rules) because that is not precise enough. It would also be incorrect to record the value as 1.920 because that implies a level of precision that does not exist with that measuring device.

Examine the three rulers below and the line being measured.

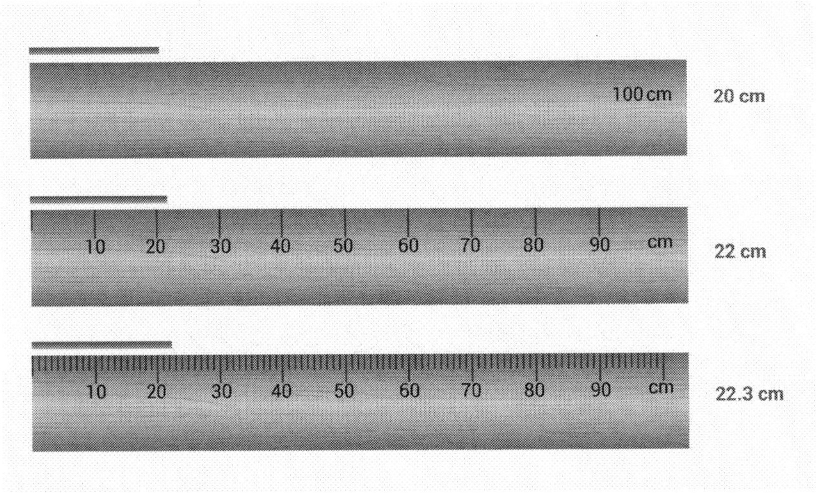

While the length of the line remains the same, the different calibrations of the rulers result in answers with a different number of significant figures. The top ruler cannot even give a level of certainty to the ones place so the appropriate number of significant figures is 1. The middle ruler does have a calibration that allows a measurement to the ones place so there are two significant figures. The bottom ruler can record a reading with 3 significant figures.

Measuring volume with a graduated cylinder is tricky because the liquid will look curved. The curve, or meniscus, is due to the adhesion of the water to the sides of the device. The bottom of the meniscus indicates the actual volume of the liquid. In this case, recording the value as 35 would be incorrect. The calibration of the graduated cylinder calls for a greater level of precision in any data taken, and the data recorded needs to reflect that. The correct value is 35.0.

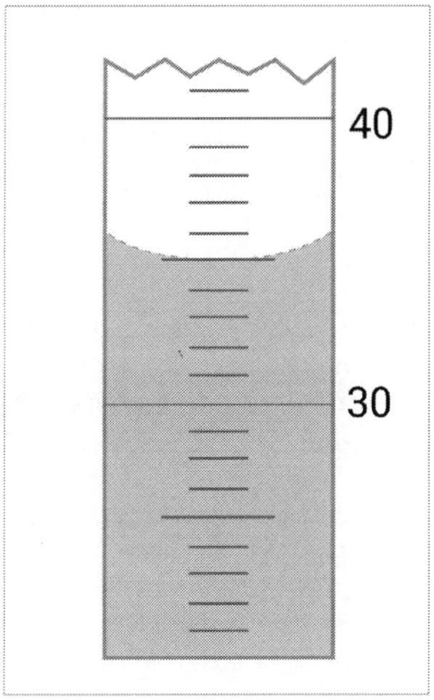

Significant figures have a few rules. Any non-zero number preceding a decimal point is significant and any trailing zeros are not significant. Zeroes between numbers are always significant. After the decimal point, any preceding zeroes are not significant while any trailing zeroes are.

- 3,570,000 has 3 significant figures.
- 4, 000, 560 has 6 significant figures.
- 0.000008320 has 4 significant figures.
- 0.00903 has 3 significant figures.

Scale

Different situations call for different units of measurement, or different scales. It is important to be mindful of scale especially when working in medicine. Anesthesia administration is weight-dependent, so it is critical to make sure that the weight is correctly labeled as either kilograms or pounds.

Kilograms or pounds, on the other hand, are inappropriate measurements for very small objects. Chemists and physicists do calculations with atoms and molecules which require a completely different unit. They use a unit called the mole which contains 6.02×10^{23} particles of a substance.

Temperature has three different scales: Kelvin, Celsius, and Fahrenheit. Kelvin is required for gas calculations, and should engineers use the wrong conversion, gas tanks could explode or be completely non-functional.

Unit Size

In understanding relationships between objects encountered in science (such as parts of the human body), it's important to roughly understand the scale of them. This is accomplished by understanding the size of different units of measurement and knowing which unit makes the most sense for measuring something. For example, you could record the distances between cities in centimeters but miles or kilometers makes the most sense. Be sure the memorize the metric prefixes discussed in the math section. That will give you an indication of how big or small a unit is.

Common Units of Measurement

There are two primary systems of measurement: the metric system and the English system. The metric system is a system of decimals, and the English system is composed of arbitrary measurements.

<u>The Metric System</u>
The Metric System was developed in France during the French Revolution in 1799, in order to distinguish itself from the British Empire and also to create a universal system of measurement. Almost every country, except three, has adopted it as their official measuring system. The physical quantities of the metric system are length, mass, volume, and time, and each has a basic unit of measurement: the meter (m), the gram (g), the liter (L), and the second (s), respectively.

Physical Quantity	Basic Unit of Measurement	Symbol
Length	meter	M
Mass	gram	G
Volume	liter	L
Time	second	S

Conversions within these measurements are easily made by multiplying or dividing by factors of ten and are indicated by the prefix attached to the unit. The most common units are listed below:

Prefix	**Symbol**	**Factor**	**Power**	**Examples**
kilo-	K	1,000	10^3	kilometer, kilogram, kiloliter, kilosecond
hecto-	H	100	10^2	hectometer, hectogram, hectoliter, hectosecond
deca-	Da	10	10^1	decameter, decagram, decaliter, decasecond
None	None	1	10^0	meter, gram, liter, second
deci-	D	0.1	10^{-1}	decimeter, decigram, deciliter, decisecond
centi-	C	0.01	10^{-2}	centimeter, centigram, centimeter, centisecond
milli-	M	0.001	10^{-3}	millimeter, milligram, milliliter, millisecond

Conversions within the Metric System

The wonderful thing about the metric system is it's incredibly easy to convert within the same unit of measurement—simply multiply or divide by ten!

Suppose a scientist wants to convert 23 meters to kilometers. The first method is to simply move the decimal point to the right or left depending on the prefix.

		LEFT			*RIGHT*	
kilo-	*hecto-*	*deca-*	*UNIT*	*deci-*	*centi-*	*milli-*
3	*2*	*1*		*1*	*2*	*3*

A decimal point for whole numbers always follows the ones unit—in this case, after the 3. To remember where the decimal is placed, imagine the number instead as 23.0.

After the decimal has been located, move it in accordance with the prefix. As shown above, *kilo* is three places to the left of the unit (meters, in this case). Therefore, the decimal is moved three units to the left:

$$23.0m \rightarrow {}^{\downarrow}\underbrace{023}_{3\,2\,1}0 \rightarrow 0.023km$$

Thus, 23 meters is equivalent to 0.023 kilometers.

Another method of conversion relies on forming an equation. To demonstrate, one should follow the steps below:

1. Write down the problem.

$$23m = ?\,km$$

2. Identify the conversion.

$$1km = 1{,}000m$$

3. Write down the conversion as a fraction.

$$\frac{1km}{1{,}000m}$$

4. Write the original unit and conversion fraction as a multiplication problem.

$$23m \times \frac{1km}{1{,}000m}$$

5. Cancel units that appear on the top and bottom.

$$23\cancel{m} \times \frac{1km}{1{,}000\cancel{m}}$$

6. Solve the equation.

$$23\cancel{m} \times \frac{1km}{1{,}000\cancel{m}} = 0.023km$$

Again, 23 meters is equivalent to 0.023 kilometers.

<u>The English System</u>

The English system of measurement was developed centuries ago using the human body as well as familiar objects. The inch, for example, was the distance between King Henry I's second and third knuckle, and the foot was distance between his heel and toes. The common units of the English system are listed below:

Length	Mass	Volume
1 foot (ft.) = 12 inches (in)	16 ounces (oz.) = 1 pound (lb.)	16 fluid oz. (fly oz.) = 1 pint
3 feet (ft.) = 1 yard (yd.)		2 pints = 1 quart
5,280 feet (ft.) = 1 mile (mi)	2,000 pounds (lb.) = 1 ton	4 quarts = 1 gallon

Conversion within the English System

Due to the arbitrary nature of the English system, one cannot simply convert one unit to another by moving decimals. The best way to convert is to use the second method—equations—listed above.

Say, for example, a scientist wanted to convert 3.6 miles into feet. The second method for this problem could be condensed into the following:

$$3.6\,\cancel{mi} \times \frac{5{,}280ft}{1\cancel{mi}} = 19{,}008ft$$

<u>Conversions Between Systems</u>

Converting units between systems can be tricky because they follow different rules. The best way is to look up an English to Metric system conversion factor and then use the second method to solve the problem. The below table lists the common conversion values:

English System	Metric System
1 inch	2.54 cm
1 foot	0.3048 m
1 yard	0.914 m
1 mile	1.609 km
1 ounce	28.35 g
1 pound	0.454 kg
1 fluid ounce	29.574 mL
1 quart	0.946 L
1 gallon	3.785 L

Suppose a scientist wanted to convert 3.6 miles to meters. The second method for this problem could be condensed into the following:

$$3.6\cancel{mi} \times \frac{1.609\cancel{km}}{1\cancel{mi}} \times \frac{1{,}000m}{1\cancel{km}} = 5{,}792m$$

Scientific Notation and Significant Figures in Collected Data

<u>Scientific Notation</u>

Scientific notation is the conversion of extremely small or large numbers into a format that is easier to comprehend and manipulate. It changes the number into a product of two separate numbers: a digit term and an exponential term.

Scientific notation = digit term x exponential term

To put a number into scientific notation, one should use the following steps:

- Move the decimal point to after the first non-zero number to find the digit number.
- Count how many places the decimal point was moved in step 1.
- Determine if the exponent is positive or negative.
- Create an exponential term using the information from steps 2 and 3.
- Combine the digit term and exponential term to get scientific notation.

For example, to put 0.0000098 into scientific notation, the decimal should be moved so that it lies between the last two numbers: 00000<u>9.8</u>. This creates the digit number:

9.8

Next, the number of places that the decimal point moved is determined; to get between the 9 and the 8, the decimal was moved six places to the right. It may be helpful to remember that a decimal moved to the right creates a negative exponent, and a decimal moved to the left creates a positive exponent. Because the decimal was moved six places to the right, the exponent is negative.

Now, the exponential term can be created by using the base 10 (this is *always* the base in scientific notation) and the number of places moved as the exponent, in this case:

$$10^{-6}$$

Finally, the digit term and the exponential term can be combined as a product. Therefore, the scientific notation for the number 0.0000098 is:

$$9.8 \times 10^{-6}$$

<u>Significant Figures</u>

Significant figures are numbers that contribute meaning to a measurement. Reporting values in significant figures reduces unnecessary numbers while increasing accuracy and minimizing confusion. For example, if a scale measures a sample to four significant figures, say, 12.56 grams, it would be inaccurate to write the number as 12.5600, because it very well may be more or less than 12.5600, but the scale cannot provide the measurement to that degree of precision. Therefore, all data must be presented with the number of figures that accurately reflect the precision of the measuring instrument.

There are rules for identifying the number of significant figures in a number:

- All non-zero digits are significant.

  For example, the number 23 has two significant figures, and the number 165.74 has five significant figures.

- Any zeros between two non-zero numbers are significant.

  For example, the number 203 has three significant figures: 2, 0, and 3.

- Leading zeros are never significant.

  For example, the value 0.000034 has two significant figures: 3 and 4.

- Trailing zeros (those following a non-zero number) are only significant after a decimal point.

  For example, the value 0.034500 has five significant figures: 3, 4, 5, 0, and 0.

- In numbers without decimals, trailing zeros may or may not be significant.

  For example, a number like 1,600 may have two or four significant figures, depending if the number is precise to the nearest unit (four) or if it's just an estimate (two).

Scientific notation is often used to reduce numbers to significant figures. In scientific notation, the exponent doesn't count as a significant figure. For example, to reduce the number 0.0000098 (which has two significant figures) into a number that only has significant figures, it can be written in scientific notation 9.8×10^{-6}, where the 9 and the 8 are the only significant numbers. It's important to notice here that the number of significant figures remains the same, but one has an unnecessary number of zeros, and the other has none.

Organizing and Presenting Data

Observations made during a scientific experiment are organized and presented as data. Data can be collected in a variety of ways, depending on the purpose of the experiment. In testing how light exposure affects plant growth, for example, the data collected would be changes in the height of the plant relative to the amount of light it received. The easiest way to organize collected data is to use a **data table.**

A data table always contains a title that relates the two variables in the experiment. Each column or row must contain the units of measurement in the heading only. See the below example (note: this is not actual data).

Plant Growth During Time Exposed to Light (130 Watts)	
Time (Hours)	Height (cm)
0	3.2
192	5.0
480	7.9
720	12.1

Data must be presented in a concise, coherent way. Most data are presented in graph form. The fundamental rule for creating a graph based on data is that the independent variable (i.e., amount of time exposed to light) is on the x-axis, and the dependent variable (i.e., height of plant) is on the y-axis.

There are many types of graphs that a person may choose to use depending on which best represents the data. These include:

- Illustrative diagrams
- Region graphs
- Line graphs
- Bar graphs
- Circle graphs
- Scatter Plots

Examples for the illustrative diagram and region graph follow on the next couple of pages, while examples of the remaining graphs can be found on pages 110-115.

The **illustrative diagram** provides a graphic representation or picture of some process. Questions may address specific details of the process depicted in the graphic. For example, "At which stage of the sliding filament theory of muscle contraction does the physical length of the fibers shorten (and contract)?"

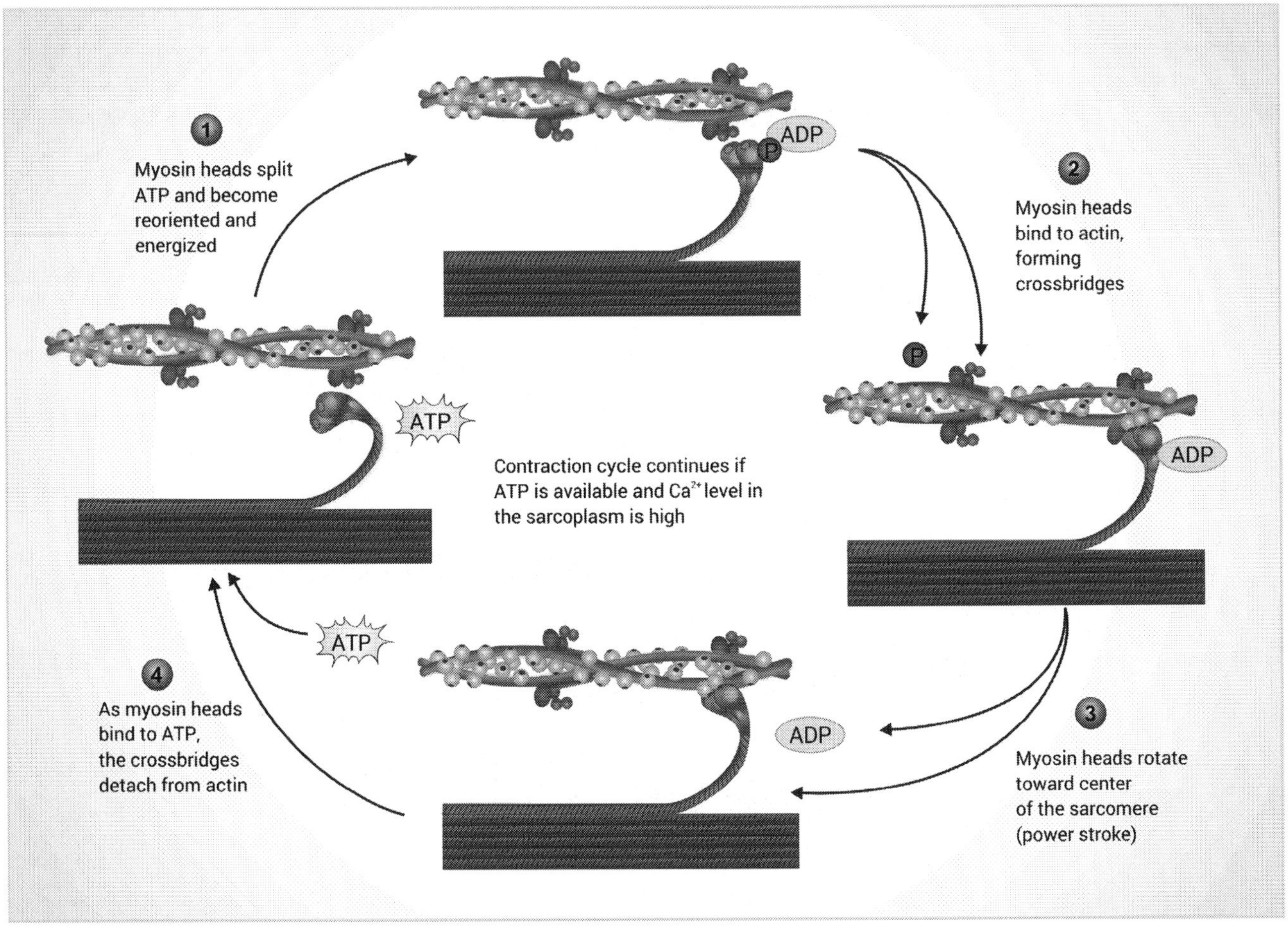

A **region graph** is a visual representation of the passage data set used to display the properties of a given substance under different conditions or at different points in time. Questions relating to this graph may ask, "According to the figure, what is the temperature range associated with liquid nitrogen?"

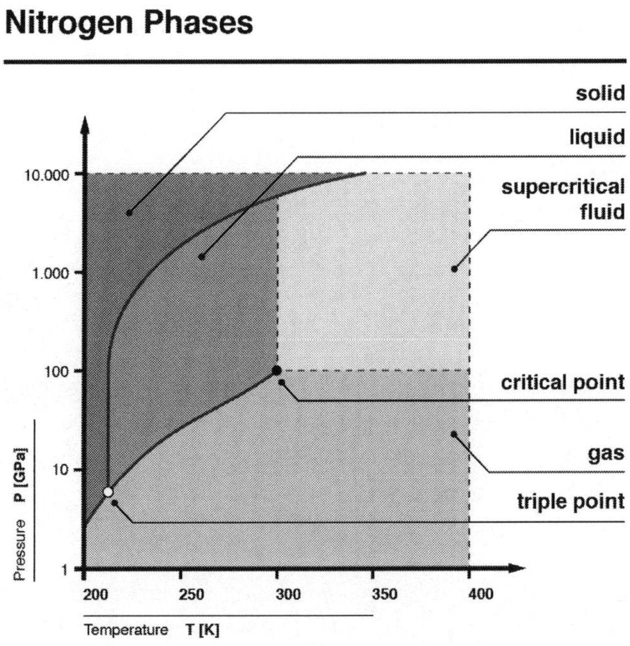

Critiquing a Scientific Explanation Using Logic and Evidence

A scientific explanation has three crucial components—a claim, evidence, and logical reasoning. A claim makes an assertion or conclusion focusing on the original question or problem. The evidence provides backing for the claim and is usually in the form of scientific data. The scientific data must be appropriate and sufficient. The scientific reasoning connects the claim and evidence and explains why the evidence supports the claim.

Scientific explanations must fit certain criteria and be supported by logic and evidence. The following represent scientific explanation criteria. The proposed explanation:

- Must be logically consistent
- Must abide by the rules of evidence
- Must report procedures and methods
- Must be open to questions and possible modification
- Must be based on historical and current scientific knowledge

The scientific method encourages the growth and communication of new information and procedures among scientists. Explanations of how the natural world shifts based on fiction, personal convictions, religious morals, mystical influences, superstitions, or authorities are not scientific and therefore irrelevant.

Scientific explanations have two fundamental characteristics. First, they should explain all scientific data and observations gleaned from experiments. Second, they should allow for predictions that can be verified with future experiments.

It's important to focus on the **empirical,** or observable/tangible evidence, in an experiment. After all, an experiment is about testing something, not theorizing about it in the abstract. Doing so helps remove a researcher's personal **bias** towards what they think or hope to be true.

Different Ways in Which Scientific Information is Presented

The purpose of scientific experimentation is to gain knowledge and share it with others, particularly in the scientific field. In scientific practice, there are structures in place that allow scientists to share new ideas and groundbreaking discoveries with other scientists and potentially the public.

The following lists some outlets that scientists use to present information:

- **Scientific journals**: Scientific journals are magazines that publish scholarly articles relating to important discoveries or new research conducted by scientists. The articles in scientific journals are peer-reviewed, which means that other experts in the scientific community review the information for fallacies or bias before the information is published. Therefore, scholarly journals publish credible information written and reviewed by experts in the field.

- **Academic conferences**: In order to formally present new developments and meet up with others in the field, scientists frequently hold and attend scientific conferences. Presenting information is usually given on a panel or from keynote speakers. Networking is an especially important part of academic conferences, as up-and-coming scientists have the opportunity to meet noteworthy experts in their field.

- **Popular media**: Communicating new research to the public is also an important way for scientists to gain acknowledgement and funding for their work. Scientists are able to reach a large audience by using newspapers, magazines, or TV stations. A few examples of popular media outlets are *National Geographic*, *The New York Times*, and *CNN*, among others.

Science Theories and Processes

Theories, models, and laws have one thing in common: *they develop on the basis of scientific evidence that has been tested and verified by multiple researchers on many different occasions*. Listed below are their exact definitions:

- **Theory:** An explanation of natural patterns or occurrences—i.e., the theory of relativity, the kinetic theory of gases, etc.

- **Model:** A representation of a natural pattern or occurrence that's difficult or impossible to experience directly, usually in the form of a picture or 3-D representation—i.e., Bohr's atomic model, the double-helix model of DNA, etc.

- **Law:** A mathematical or concise description of a pattern or occurrence in the observable universe—i.e., Newton's law of gravity, the laws of thermodynamics, etc.

The terms *theory, model,* and *law* are often used interchangeably in the sciences, although there's an essential difference: theories and models are used to explain *how* and *why* something happens, while laws describe exactly *what* happens. A common misconception is that theories develop into laws. But theories and models never become laws because they inherently describe different things.

Type	Function	Examples
Theory	To explain how and why something happens	Einstein's Theory of Special Relativity The Big Bang Theory
Model	To represent how and why something happens	A graphical model or drawing of an atom
Laws	To describe exactly what happens	$E = mc^2$ $F = ma$ $PV = nRT$

In order to ensure that scientific theories are consistent, scientists continually gather information and evidence on existing theories to improve their accuracy.

Claims and Evidence in Science

Deciding Whether Conclusions are Supported by Data

Drawing conclusions is the process of analyzing patterns in data and determining whether the relationship is **causal**, meaning that one variable is the cause of the change in the other. There are many correlations that aren't casual, such as a city where alcohol sales increase as crime increases. Although there's a positive correlation between the two, crime may not be the factor that causes an increase in alcohol sales. There could be other factors, such as an increase in unemployment, which increases both alcohol sales and crime rates. Although crime and alcohol sales are positively correlated, they aren't causally correlated.

For this reason, it's important for scientists to carefully design their experiments with all the appropriate constants to ensure that the relationships are causal. If a relationship is determined to be causal by isolating the variables from all other factors, only then can conclusions be drawn based on data. In the plant growth experiment, the conclusion is that light affects plant growth, because the data shows they are causally correlated since the two variables were entirely isolated.

Finding Evidence that Supports a Finding

Science is one of the most objective, straightforward fields of study. Thus, it is no surprise that scientists and science articles are focused on **evidence**. When reading science passages, test takers are sometimes asked to find supporting evidence that reinforces a particular finding. A **finding** in science is a result of the investigation; it is what scientists find out. The majority of science passages tend to avoid opinions; instead, they focus on facts. Although no results are infallible just because the texts are scientific, most results are quantified. Quantified results mean they are expressed in numbers or measurements. Thus, when in doubt, go straight to the data, or numbers, that are offered. Sometimes data is embedded in the text; other times it appears in charts, tables, or graphs. These tools use numbers to demonstrate the patterns discussed in scientific texts, and they help readers to visualize concrete patterns. In order to find evidence to support a finding in scientific passage, all test takers should try collecting and analyzing the relevant data offered. Regardless of whether the data is coming from the text or a graph, it is helpful when making conclusions.

The following steps are helpful for identifying evidence that supports a finding in a science passage:

- Apply critical analysis and critical thinking by asking the right questions.
- Determine the weight of the information by figuring out its relevance.
- Identify trends in the numbers.
- Make inferences.
- Determine the most appropriate methods for either quantifying or communicating inferences.

<u>Making Sense of Information that Differs Between Various Science Sources</u>

Science is often a process of checks and balances, and students are expected to carry out this process of checks and balances as they analyze and compare information that differs between various science sources. Science demands a high degree of communication, which, in turn, demands a high degree of scientific literacy and numeracy. Students must be prepared to analyze the different data and written conclusions of various texts. Contrary to popular belief, science is not an authoritarian field—scientific worldviews and inquiries can be wrong. It is more fruitful to think of science as a living library that is shaped by the complex activities carried out by different groups in different places. This living library is filled with ideas that are shaped by various sources and methods of research. The explanations, inferences, and discussions carried out by scientists are filled with facts that may be flawed or biased. Science, like any other field, cannot completely escape bias. Even though science is meant to be objective, its findings can still lend themselves to biases.

Thus, it is important for students to get in the practice of not only making sense of information that differs between various science sources, but also to begin synthesizing this information into a unique worldview. The peer review process is also necessary to ensure checks and balances within the scientific field. Pay close attention to the ways in which each scientist uses specific words or data to back their overall conclusions.

Below are some key reasons why data and interpretations can differ:

- Historical bias
- Cultural bias
- Interpretation or personal bias
- Lack of implementation and data collection fidelity
- Different data collection approaches
- Different data collection and data analysis tools
- Weak hypotheses
- Compounding variables
- Failure to recognize certain variables
- User error
- Changes in the environment between two studies
- Computation or statistical errors
- Interpretive blind spots
- Lack of understanding of context or environment

<u>Appropriate Samples</u>

When conducting a scientific experiment, once a hypothesis is conceived, it is important for scientists and students to adequately test their hypothesis. In order to test a hypothesis, scientists and students must analyze a particular **sample** of the larger population. Every sample must be composed of two comparable groups: a treatment (or experimental) group and a control group. Control groups can be used to acquire baseline measurements for analysis. When conducting the experiment, the treatment groups are the ones affected by experimental manipulation. From this sample, scientists and students can compare relevant data from each group to test their hypothesis.

If the experimental manipulation is effective, then scientists and students notice evident differences in the data collected from each sub-sample. In some cases, when scientists are merely carrying out observations, they may not split the sample into any subgroups but instead try to create new sub-samples through observation and data collection. Regardless of the approach, these kinds of samples, whether physical or observatory, are necessary for conducting scientific research. For the TEAS, test takers should be aware of how scientists use these samples, how they display their results, and how they compare samples through qualitative and quantitative data. Additionally, they should be prepared to extrapolate findings from the sample, comparing it to the larger population or other subpopulations. This extrapolation helps test takers better understand whether the sample is a good representation of the larger population or comparative to other subpopulations.

Relationships Among Events, Objects, and Processes

Cause and Effect

A **cause** is like a stimulus in the nervous system in that it results in a response. An **effect** is what happens because of the stimulus. The cause is *why* something happens, and the effect is *what* actually happens:

Cause	Effect
I forgot to wear shoes.	I stubbed my toe.
I forgot to practice piano.	I messed up at the recital and was embarrassed.
I studied for my math exam.	I got a 100.

While it is typical for there to be a single cause and a single effect as noted above, there are many situations that call for a cause to have many effects, such as exploring the effects of a certain event. In exploring the effects of exercise, for example, the cause is exercise. Effects are stress relief, increased energy, and weight loss.

There is the possibility that one effect can have many causes. For example, type 2 diabetes is not a one-cause disease, but rather a combination of exercise habits, diet, and genetics.

A cause-and-effect chain is when an effect then causes another effect. For example, the initial cause could be when a young gentleman played video games instead of studying for a biology test, which sets off a chain reaction of effects.

- 1st effect: He failed the test.
- 2nd effect: He got in trouble with his parents.
- 3rd effect: He couldn't hang out with his friends on Friday night.
- 4th effect: He had a boring night.

Using Evidence to Draw Conclusions or Make Predictions

Making Conclusions Based on Data

In general, there are two types of data: qualitative and quantitative. Science passages may contain both, but simply put, **quantitative** data is reflected numerically and qualitative is not. **Qualitative** data is based on its qualities. In other words, qualitative data tends to present information more in subjective generalities (for example, relating to size or appearance). Quantitative data is based on numerical findings such as percentages. Quantitative data will be described in numerical terms. While both types of data are valid, the test taker will more likely be faced with having to interpret quantitative data through one or more graphic(s), and then be required to answer questions regarding the numerical data. A test taker should take the time to learn the skills it takes to interpret quantitative data so that they can make sound conclusions.

A **line graph** presents quantitative data on both horizontal (side to side) and vertical (up and down) axes:

Cell Phone Use in Kiteville, 2000-2006

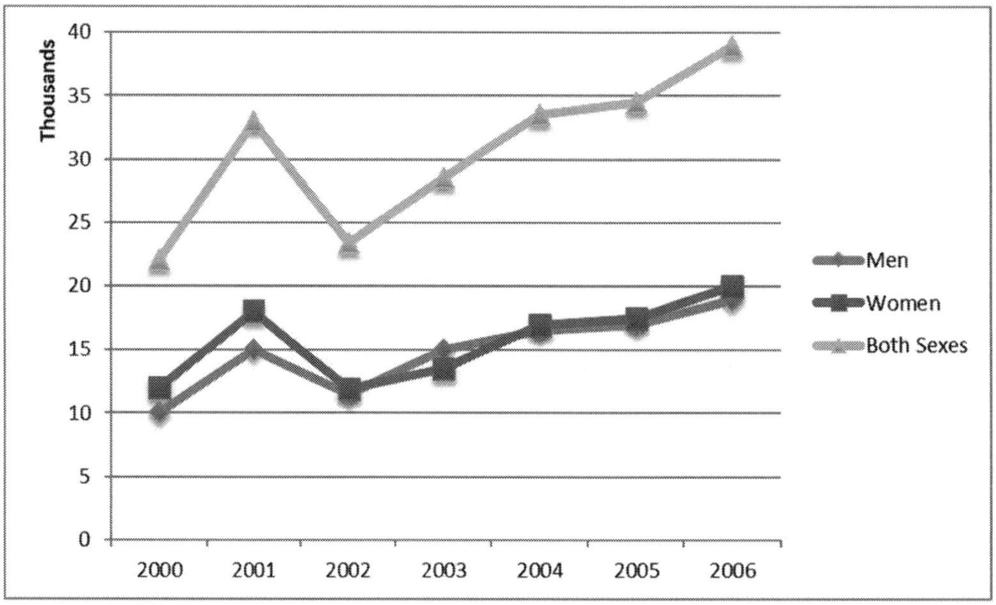

It requires the test taker to examine information across varying data points. The horizontal axis represents the passage data sets that are compared over time. The vertical axis is the scale for measurement of that data. The scale points are equidistant from one another. When reading a line graph, a test taker should pay attention to any headings, as these indicate a title for the data it contains. In the above example, the test taker can anticipate the line graph contains numerical data regarding the use of cellphones during a certain time period. From there, a test taker should carefully read any outlying words or phrases that will help determine the meaning of data within the horizontal and vertical axes. In this example, the vertical axis displays the total number of people in increments of 5,000. Horizontally, the graph displays yearly markers, and the reader can assume the data presented accounts for a full calendar year. In addition, the line graph also defines its data points by shapes. Some data points represent the number of men. Some data points represent the number of women, and a third type of data point represents the number of both sexes combined.

A test taker may be asked to read and interpret the graph's data, then answer questions about it. For example, the test may ask, *In which year did men seem to decrease cellphone use?* then require the test taker to select the correct answer. Similarly, the test taker may encounter a question such as *Which year yielded the highest number of cellphone users overall?* The test taker should be able to identify the correct answer as 2006.

A **bar graph** presents quantitative data through the use of lines or rectangles. The height and length of these lines or rectangles corresponds to contain numerical data. The data presented may represent information over time, showing shaded data over time or over other defined parameters. A bar graph will also utilize horizontal and vertical axes. TEAS bar graphs will be printed in black and white. Data may be oriented vertically or horizontally.

An example of a vertical bar graph is as follows:

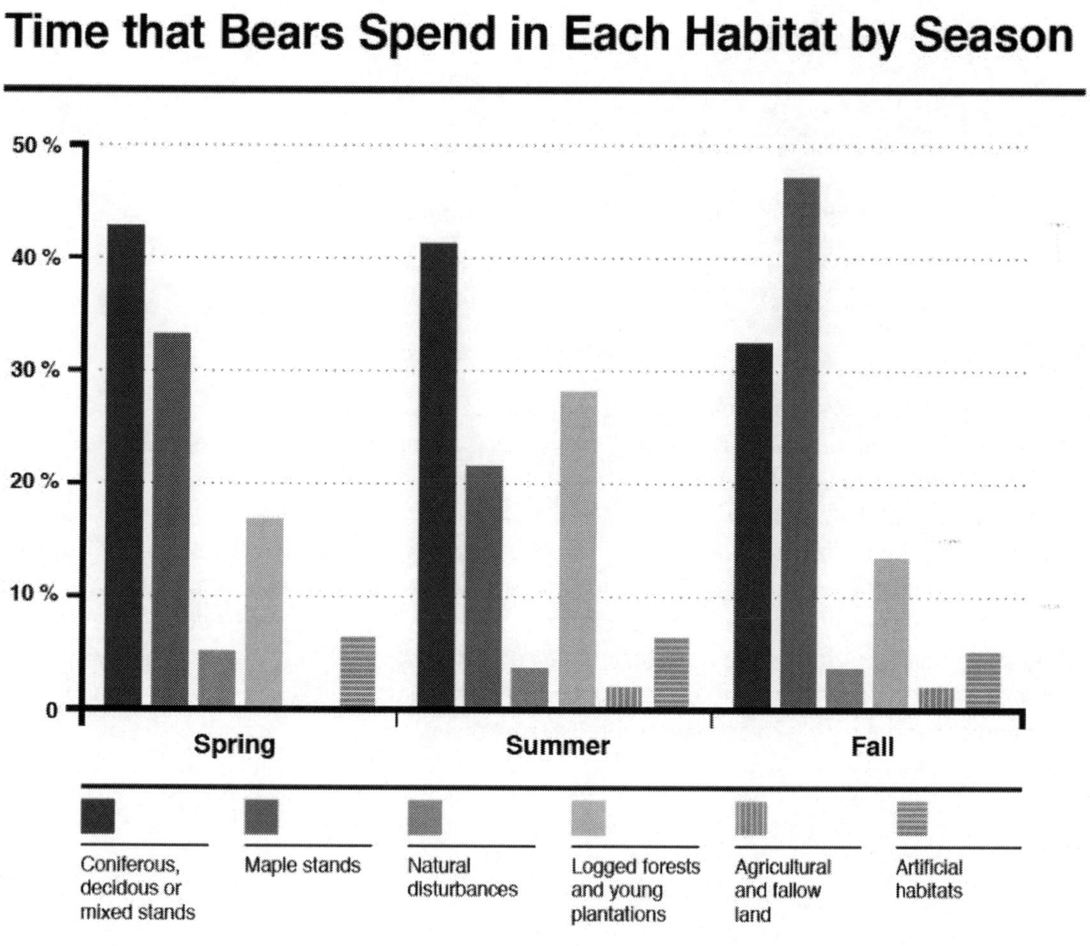

Here is an example of a horizontal bar graph:

Population Growth in Major U.S. Cities

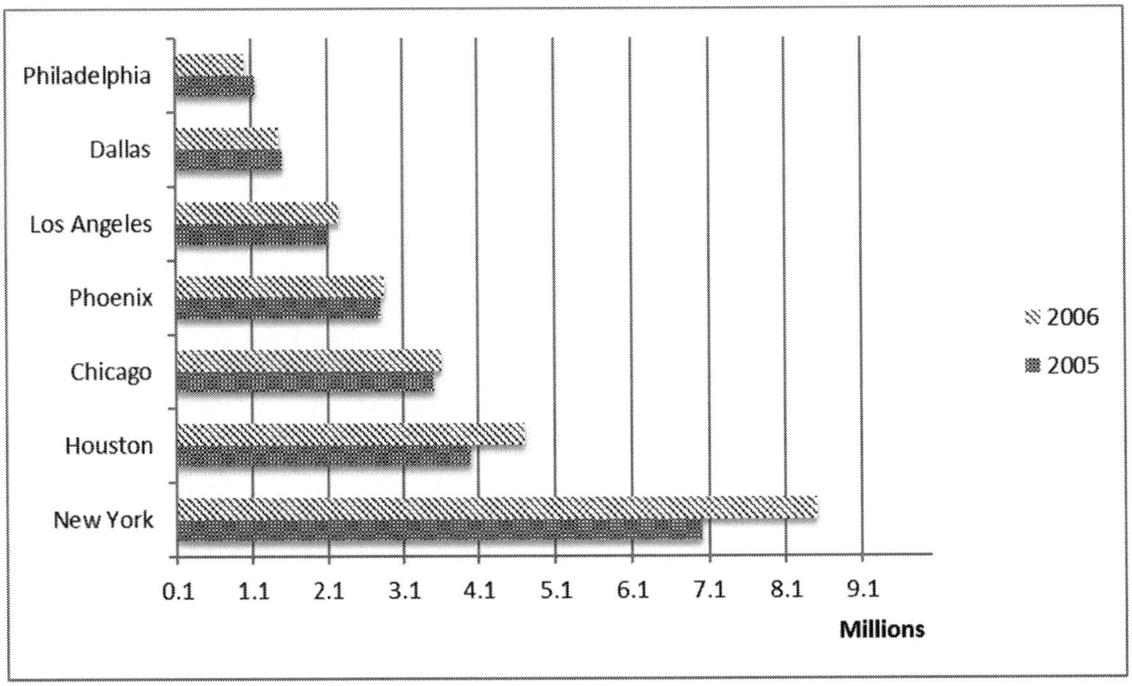

Reading the data in a bar graph is similar to the skills needed to read a line graph. The test taker should read and comprehend all heading information, as well as information provided along the horizontal and vertical axes. Note that the graph pertains to the population of some major U.S. cities. The "values" of these cities can be found along the left side of the graph, along the vertical axis. The population values can be found along the horizontal axes. Notice how the graph uses shaded bars to depict the change in population over time, as the heading indicates. Therefore, when the test taker is asked a question such as, *Which major U.S. city experienced the greatest amount of population growth during the depicted two year cycle,* the reader should be able to determine a correct answer of New York. It is important to pay particular attention to color, length, data points, and both axes, as well as any outlying header information in order to be able to answer graph-like test questions.

A **circle graph** presents quantitative data in the form of a circle (also sometimes referred to as a **pie chart**) and is used when the data sum to 100%. The same principles apply: the test taker should look for numerical data within the confines of the circle itself but also note any outlying information that may be included in a header, footer, or to the side of the circle. A circle graph will not depict horizontal or vertical axis information but will instead rely on the reader's ability to visually take note of segmented circle pieces and apply information accordingly.

An example of a circle graph is as follows:

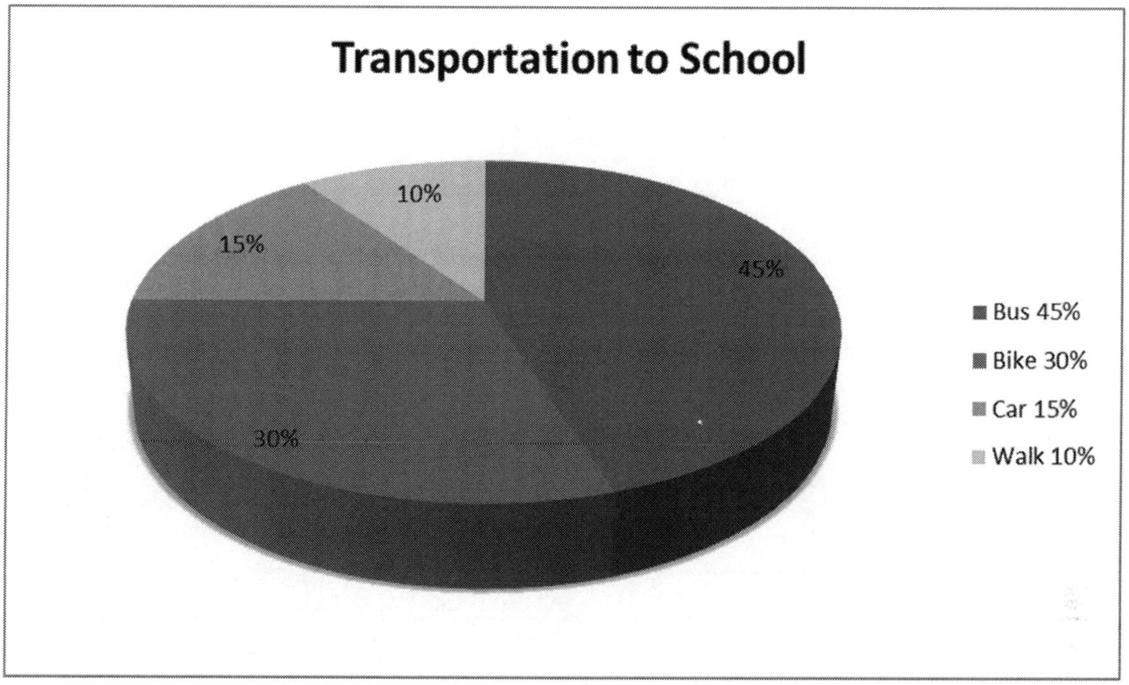

Notice the heading "Transportation to School." This should indicate to the test taker that the topic of the circle graph is how people traditionally get to school. To the right of the graph, the reader should comprehend that the data percentages contained within it directly correspond to the method of transportation. In this graph, the data is represented through the use shades and pattern. Each transportation method has its own shade. For example, if the test taker was then asked, *Which method of school transportation is most widely utilized,* the reader should be able to identify school bus as the correct answer.

Be wary of test questions that ask test takers to draw conclusions based on information that is not present. For example, it is not possible to determine, given the parameters of this circle graph, whether the population presented is of a particular gender or ethnic group. This graph does not represent data from a particular city or school district. It does not distinguish between student grade levels and, although the reader could infer that the typical student must be of driving age if cars are included, this is not necessarily the case. Elementary school students may rely on parents or others to drive them by personal methods. Therefore, do not read too much into data that is not presented. Only rely on the quantitative data that is presented in order to answer questions.

A **scatter plot** or **scatter diagram** is a graph that depicts quantitative data across plotted points. It will involve at least two sets of data. It will also involve horizontal and vertical axes.

Scatter plots provide a visual representation of the passage data along the x- and y-axes:

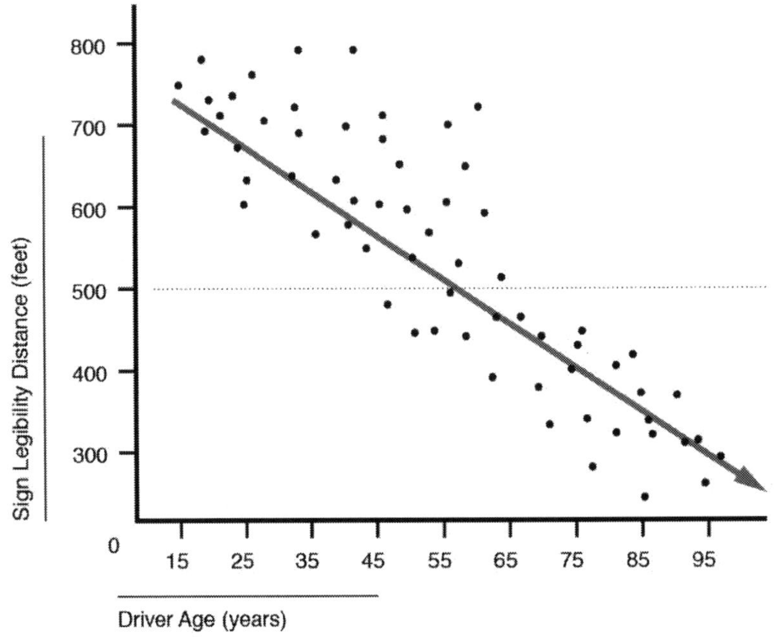

The skills needed to address a scatter plot are essentially the same as in other graph examples. Note any topic headings, as well as horizontal or vertical axis information. In the sample above, the reader can determine the data addresses a driver's ability to correctly and legibly read road signs as related to their age. Again, note the information that is absent. The test taker is not given the data to assess a time period, location, or driver gender. It simply requires the reader to note an approximate age to the ability to correctly identify road signs from a distance measured in feet. It is important to note that correlation doesn't equal causation. The relationship may be linear, curvilinear, positive, negative, inverse, or there may be no relationship. Notice that the overall graph also displays a trend. In this case, the data indicates a negative one and possibly supports the hypothesis that as a driver ages, their ability to correctly read a road sign at over 500 feet tends to decline over time. If the test taker were to be asked, *At what approximation in feet does a sixteen-year-old driver correctly see and read a street sign,* the answer would be the option closest to 500 feet.

Reading and examining scientific data in excerpts involves all of a reader's contextual reading, data interpretation, drawing logical conclusions based only on the information presented, and their application of critical thinking skills across a set of interpretive questions. Thorough comprehension and attention to detail is necessary to achieve test success.

Making Predictions Based on Data

Science is amazing in that it actually allows people to predict the future and see into the past with a certain degree of accuracy. Using numerical correlations created from quantitative data, one can see in a general way what will happen to *y* when something happens to *x*.

The best way to get a useful overview of quantitative data to facilitate predictions is to use a scatter plot, which plots each data point individually. As shown above, there may be slight fluctuations from the correlation line, so one may not be able to predict what happens with *every* change, but he or she will be able to have a general idea of what is going to happen to *y* with a change in *x*. To demonstrate, the graph with a line of best fit created from the plant growth experiment is below.

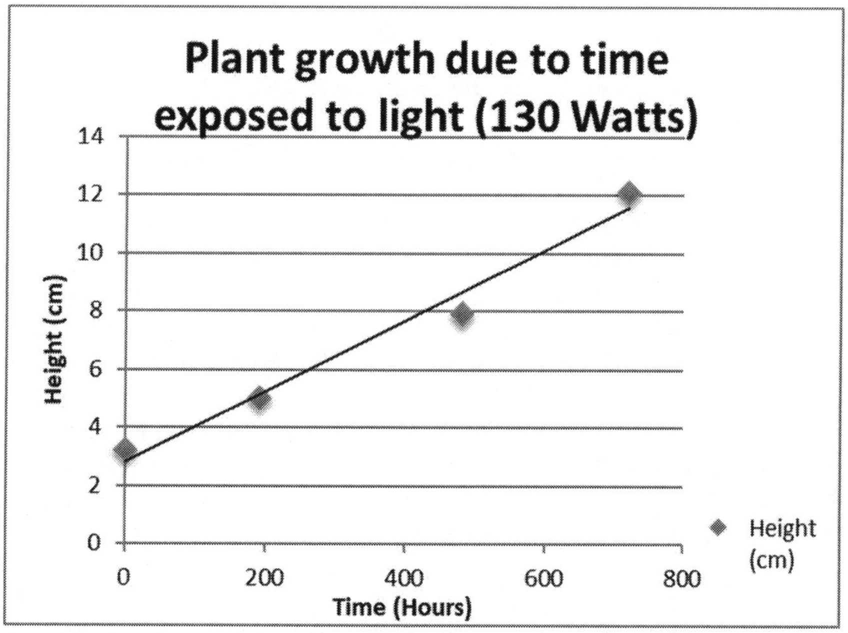

Using the trend line within the data, one can estimate what will happen to plant growth at a given length of time exposed to light. For example, it can be estimated that with 700 hours of time, the plant is expected to grow to a height of about 11 cm. The plant may not grow to exactly 11 cm, but it will likely grow to about that height based on previous data. This process allows scientists to draw conclusions based on data.

Science Formulas

Scientific inquiry includes the fields of chemistry and physics. It incorporates mathematical problems, which are described with science formulas. Science formulas cannot be used haphazardly. In fact, these science formulas have very specific standards for use. These formulas stand at the intersection of mathematics and science. And, like math, science extracts all of its meaning through its specificity. On any standardized science test, the biggest challenge is knowing when to employ a particular science formula. All formulas are provided on the examination. Thus, the test taker does not need to memorize these formulas; they just need to use the formulas correctly.

Take the dynamics of motion as an example. In science, dynamics is the study of the relationship between motion and the forces affecting motion. In science, there is a formula that is universally used for force: F = m/a. This equation is shorthand for force (F) equals mass (m) divided by (/) acceleration (a). This equation is used to describe the relationship with three components of science: force, mass, and acceleration. For instance, an object remains at rest unless the force (F) is strong enough to move the mass (m) and cause the object to accelerate (a).

However, the equation F = m/a would NOT be helpful for figuring out the weight of an object. The weight of an object demands another formula: $W = m \times g$. This formula is shorthand for weight (W) equals mass (m) times (×) gravity (g). Force and weight have their own scientific formulas. The formulas do not need to be memorized, but every test taker needs to know *when* and *how* to use a particular formula. Otherwise, their results will be incorrect.

Using Statistics to Describe Science Data

In an experiment, variables are the key to analyzing data, especially when data is in a graph or table. Variables can represent anything, including objects, conditions, events, and amounts of time.

Covariance is a general term referring to how two variables move in relation to each other. Take for example an employee that gets paid by the hour. For them, hours worked and total pay have a positive covariance. As hours worked increases, so does pay.

Constant variables remain unchanged by the scientist across all trials. Because they are held constant for all groups in an experiment, they aren't being measured in the experiment, and they are usually ignored. Constants can either be controlled by the scientist directly like the nutrition, water, and sunlight given to plants, or they can be selected by the scientist specifically for an experiment like using a certain animal species or choosing to investigate only people of a certain age group.

Independent variables are also controlled by the scientist, but they are the same only for each group or trial in the experiment. Each group might be composed of students that all have the same color of car or each trial may be run on different soda brands. The independent variable of an experiment is what is being indirectly tested because it causes change in the dependent variables.

Dependent variables experience change caused by the independent variable and are what is being measured or observed. For example, college acceptance rates could be a dependent variable of an experiment that sorted a large sample of high school students by an independent variable such as test scores. In this experiment, the scientist groups the high school students by the independent variable (test scores) to see how it affects the dependent variable (their college acceptance rates).

Note that most variables can be held constant in one experiment but independent or dependent in another. For example, when testing how well a fertilizer aids plant growth, its amount of sunlight should be held constant for each group of plants, but if the experiment is being done to determine the proper amount of sunlight a plant should have, the amount of sunlight is an independent variable because it is necessarily changed for each group of plants.

The most common relationship examined in an experiment is between two variables (independent and dependent), most often referred to as *x* and *y*. The independent variable (*x*) is displayed on the horizontal axis of a coordinate plane, and the dependent variable (*y*) is displayed on the vertical axis.

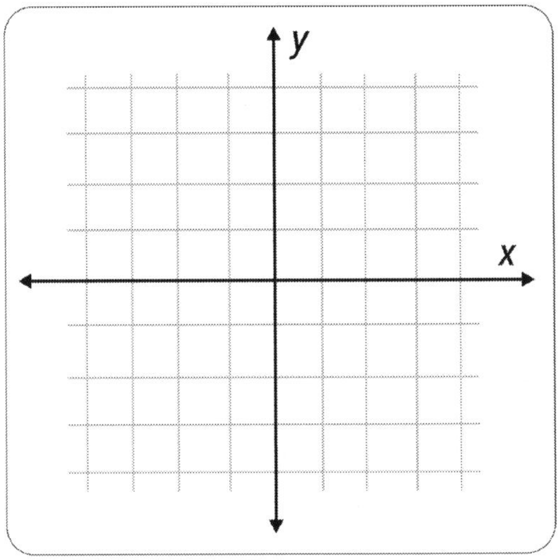

The placement of the variables in this way provides a visual representation of what happens to *y* when *x* is manipulated. In analyzing trends, *x* is used to predict *y*, and since *y* is the result of *x*, then *x* comes before *y* in time. For example, in the experiment on plant growth, the hours the plant was exposed to light had to happen before growth could occur.

When analyzing the relationship between the variables, scientists will consider the following questions:

- Does *y* increase or decrease with *x*, or does it do both?
- If it increases or decreases, how fast does it change?
- Does *y* stay steady through certain values of *x*, or does it jump dramatically from one value to the other?
- Is there a strong relationship? If given a value of *x*, can one predict what will happen to *y*?

If, in general, *y* increases as *x* increases, or *y* decreases and *x* decreases, it is known as a **positive correlation**. The data from the plant experiment show a positive correlation—as time exposed to light (*x*) increases, plant growth (*y*) increases. If the variables trend in the opposite direction of each other—that is, if *y* increases as *x* decreases, or vice versa—it is called a **negative correlation**. If there doesn't seem to be any visible pattern to the relationship, it is referred to as **no** or **zero correlation**.

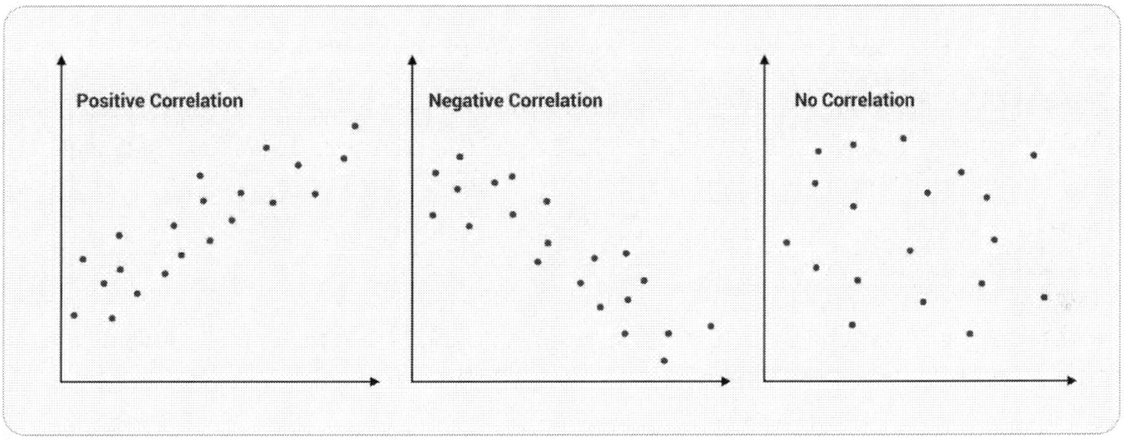

Experiments that show positive or negative correlation within their data indicate that the variables are related. This allows scientists to make predictions based on the data.

The center of a set of data (statistical values) can be represented by its mean, median, or mode. These are sometimes referred to as measures of central tendency.

Mean

The first property that can be defined for this set of data is the **mean**. This is the same as average. To find the mean, add up all the data points, then divide by the total number of data points. For example, suppose that in a class of 10 students, the scores on a test were 50, 60, 65, 65, 75, 80, 85, 85, 90, 100. Therefore, the average test score will be:

$$\frac{50 + 60 + 65 + 65 + 75 + 80 + 85 + 85 + 90 + 100}{10} = 75.5$$

The mean is a useful number if the distribution of data is normal (more on this later), which roughly means that the frequency of different outcomes has a single peak and is roughly equally distributed on both sides of that peak. However, it is less useful in some cases where the data might be split or where there are some **outliers.** Outliers are data points that are far from the rest of the data. For example, suppose there are 10 executives and 90 employees at a company. The executives make \$1000 per hour, and the employees make \$10 per hour.

Therefore, the average pay rate will be:

$$\frac{\$1000 \times 11 + \$10 \times 90}{100} = \$119 \; per\ hour$$

In this case, this average is not very descriptive since it's not close to the actual pay of the executives *or* the employees.

Median

Another useful measurement is the **median**. In a data set, the median is the point in the middle. The middle refers to the point where half the data comes before it and half comes after, when the data is recorded in numerical order. For instance, these are the speeds of the fastball of a pitcher during the last inning that he pitched (in order from least to greatest):

90, 92, 93, 93, 95, 96, 97, 97, 97

There are nine total numbers, so the middle or **median** number is the 5th one, which is 95.

In cases where the number of data points is an even number, then the average of the two middle points is taken. In the previous example of test scores, the two middle points are 75 and 80. Since there is no single point, the average of these two scores needs to be found. The average is:

$$\frac{75 + 80}{2} = 77.5$$

The median is generally a good value to use if there are a few outliers in the data. It prevents those outliers from affecting the "middle" value as much as when using the mean.

Since an outlier is a data point that is far from most of the other data points in a data set, this means an outlier also is any point that is far from the median of the data set. The outliers can have a substantial effect on the mean of a data set, but usually do not change the median or mode, or do not change them by a large quantity. For example, consider the data set (3, 5, 6, 6, 6, 8). This has a median of 6 and a mode of 6, with a mean of $\frac{34}{6} \approx 5.67$. Now, suppose a new data point of 1000 is added so that the data set is now (3, 5, 6, 6, 6, 8, 1000). This does not change the median or mode, which are both still 6. However, the average is now $\frac{1034}{7}$, which is approximately 147.7. In this case, the median and mode will be better descriptions for most of the data points.

The reason for outliers in a given data set is a complicated problem. It is sometimes the result of an error by the experimenter, but often they are perfectly valid data points that must be taken into consideration.

Mode

One additional measure to define for X is the **mode.** This is the data point that appears most frequently. If two or more data points all tie for the most frequent appearance, then each of them is considered a mode. In the case of the test scores, where the numbers were 50, 60, 65, 65, 75, 80, 85, 85, 90, 100, there are two modes: 65 and 85.

Science Vocabulary, Terms, and Phrases

Every field of study has its own specialized vocabulary, terms, and phrases. However, the field of science has one of the most unique living libraries when it comes to its specialized vocabulary, terms, and phrases. Many people mistakenly believe science is just about mathematical formulas. But scientific literacy is essential to becoming a marketable employee and an active global citizen. All students of science must have a basic understanding of science vocabulary, terms, and phrases in order to shape their scientific worldview, drive their scientific inquiry, and enhance their scientific enterprise.

Every person has the ability to formulate their own scientific worldview. Science is not only highly complex, but it is also tentative in nature—humans make educated guesses about the information and data they collect. This information and data change over time, and, as a result, new vocabulary, terms, and phrases are constantly shadowing new discoveries, data analysis, and findings. Scientific inquiry may begin with simple questions, or hypotheses, but it may end with new concepts and terminology. The more questions scientists ask, the more necessary it is to apply words to both the preconceived questions and logical answers. Scientific inquiry, in many ways, is a combination of logic and imagination, and scientific imagination is usually the force driving the creation of new specialized vocabulary, terms, and phrases. Scientific enterprise is a larger process of checks and balances, which allows scientific findings to be debated through intellectual dialogue. Words are an important component of the peer review process of scientific enterprise. Theories and findings always come into question, and students are expected to not only comprehend the words of passages, but also question their findings through intertextual analysis and data analysis.

Below are a few tables of commonly used scientific terms and their definitions:

Earth Science Terms
Carbon cycle: The series of biochemical processes that transfer carbon, the main component of many biological compounds and minerals, into different reservoirs of the Earth's environment; these reservoirs include the Earth's interior, its geological sediments, its oceans and waterways, its terrestrial biosphere, and its atmosphere.
Climate: The weather in a particular location averaged over a period of time.
Equator: The line separating the Northern and Southern Hemispheres.
Hemisphere: Literally "half a sphere"; the earth is divided into two pairs of hemispheres (Northern and Southern, Eastern and Western).
Orbit: The path of a celestial body in its revolution about another.
Prime meridian: A human-made geographic line that runs South-to-North and separates the globe into two: the Eastern and Western Hemispheres.
Revolve: Turn on or around an axis.
Seismic: Caused by an earthquake or geological vibration.
Solar system: The Sun and the celestial bodies (for example, Earth and the other planets) that orbit it.
Tectonic: Pertaining to the structure or movement of the Earth's crust.
Volcano: A geological feature composed of a mountain or crevice that spews hot gases and magma into the atmosphere when agitated.
Water: The liquid that sustains life; a colorless, odorless liquid that helps shape geography (for example, rivers, lakes, streams, oceans, glaciers) and biological functions (for example, hydration, blood flow).
Weather: Atmospheric conditions such as temperature and precipitation.
Wind: Air moving from high pressure to low pressure.

Chemistry Terms
Acid: a corrosive chemical substance that has the ability to neutralize alkali; opposite of a base on the pH scale.
Atom: The microscopic building blocks of matter and the foundations of chemistry; every solid, liquid, and gas is composed of atoms.

Atomic mass: Typically expressed in atomic mass units, atomic mass is mass of an atom, which can be figured out by adding protons and neutrons to find the mass number.

Base: Bases are substances that are the opposite of acids on the pH scale; when mixed with acids they create a chemical reaction that produces salts.

Boiling point: A point at which a liquid reaches a temperature that allows it to transform in a vapor; the normal boiling point of water is 100 degrees Celsius.

Oxidation: The process or result of oxidizing or being oxidized.

Chemical formula: An arrangement of elemental symbols and subscripts that convey the structure of a compound.

Chemical reaction: A process that transforms the molecular or ionic composition of a substance.

Condensation: The conversion of a vapor or gas to a liquid.

Conductor: Any material that can transmit sound, heat, or electricity; normally used to describe metals and other materials that serve as conduits for electricity.

Covalent bond: The sharing of electrons between two atoms; sometimes called a molecular bond.

Decompose: (With reference to a chemical compound) break down or cause to break down into component elements or simpler constituents.

Electron: Subatomic particles that carry negative charge; carries a charge that is opposite of a proton.

Ion: Atoms or molecules that are not neutral in their charge; they are positively or negatively charged.

Ionic bond: A type of ionic cohesion that occurs what two oppositely charged atoms or molecules attract to one another.

Liquid: A substance that is neither a solid nor a gas, but flows freely in a fluid-like state that is akin to oil or water.

Radioactivity: A process by which atoms become unstable and emit radiation; an unstable, disintegrating atom losing energy.

Gas: An air-like state of matter that has traits that are distinguishable from solids and liquids; gases are known for their ability to expand rapidly throughout free space because of their air-like qualities; in fact, air itself is a combination of gases.

Salt: A substance created from a mixture of a base and an acid.

Solution: A liquid combination of a solute (a chemical that dissolves in a solution) and a solvent (a chemical that other substances dissolve into)

Physics Terms

Acceleration: Acceleration in physics can be found by dividing the change in velocity by the change in time; it is the rate of the overall change in velocity as analyzed in relationship to the overall change in time.

Activation energy: The amount of energy that is required for a chemical reaction to be initiated; normally quantified by a measurement known as Joules (J).

Amplitude: The peak strength of a vibration or wave, which is measured by its repetition over a single period.

Density: The degree of compactness of a substance.

Electromagnetic spectrum: The spectrum that includes all different types of electromagnetic radiation, including radio waves, infrared waves, visible light waves, ultraviolet waves, X-rays, and Gamma rays

Energy: The scientific name describing the phenomenon that produces heat, light, or motion in an object; in biology, it is produced by cells and the body in order to fuel human functions.

Frequency: The number of occurrences within a given period of time.

Friction: The resistance when a body is moved in contact with another.

Gravity: Discovered by Sir Isaac Newton, it is an invisible force that draws two objects or celestial bodies close together; this force is displayed in space when celestial objects orbit around one another.

Momentum: Used by scientists to quantify the motion object.

Resistance: The impeding, slowing, or stopping effect exerted by one material thing on another.

Voltage: The electrical potential between two items, expressed in a measurement known as volts.

Volume: The amount of space that a three-dimensional geometrical object (whether solid, liquid, or gas) occupies.

General Science Terms

Atmosphere: In science, it is the combinations of gases that contribute to the mass and shape of a celestial body, particularly the planet Earth.

Cause: The beginning of a chain reaction or the root an event; a phenomenological starting point that produces an effect.

Conclusion: A definitive decision, assumption, or statement that is rooted in logic.

Context: In science, it is the places, perspectives, and circumstances that helped influence a particular experiment or hypothesis.

Effect: An event, process, or phenomenon that results from another event, process, or phenomenon (which is known as a cause).

Experiment: A scientific procedure that helps test hypotheses by carrying out the scientific method.

Evaluate: To assess the nature, content, necessity, or effectiveness of a particular fact, theory, or philosophy.

Evidence: The details or facts needed to prove a hypothesis, theory, case, or argument.

Fact: Information that can be substantiated as true, using science or accepted theory; information that can be cross-referenced with others' understanding of reality.

Hypothesis: A postulation or prediction that is assessed via the scientific method.

Inference: An implicit conclusion that can be reached by combining the details of a document or experience with logically sound background knowledge.

Implication: Something that is inferred.

Generalization: The process of abstracting common properties of instances.

Research: The process of consolidating information, ideas, and theories in a systematized fashion (typically via writing books, articles, or reports).

Scientific method: The consolidation of ideas, observations, experiments, and theories, using tested hypotheses as a guide for understanding the world.

Scientist: A researcher or practitioner devoted to the study of science, which includes: physics, Earth sciences, chemistry, and biology.

Theory: A worldview, or a system of philosophies, ideas, perceptions, and conjectures, that tries to make sense of reality.

Understanding Symbols, Terms, and Phrases in Science

Much like in mathematics, scientific texts rely on symbols to convey certain messages. These symbols may be in a text, or they may be in graphs, charts, tables, or pictures. These symbols are key components of evaluating data and communicating results.

In physics-based passages, students may encounter symbols and variables that are particular to that field of science. For instance, in studies focusing on electricity and magnetism, students are likely to encounter a symbol that looks like this: λ. This symbol is the Greek letter "lambda." Scientists uses the λ symbol in order to communicate or compute a wavelength, measured in meters, of the electromagnetic spectrum. Scientists might also use a symbol like this: ƒ. This is a symbol that represents the frequency, measured in hertz, of the electromagnetic spectrum.

In chemistry, symbols are also used on a regular basis. For instance, when discussing temperature, whether in Celsius or Fahrenheit, scientists often use the following symbol: °. Instead of saying 35 degrees, a scientist might just present the finding as 35°. Chemistry also uses symbols for the periodic table. For example, mercury is represented on the periodic table by using the following symbol: Hg. All elements have their own unique symbols.

In most cases, test takers have access to a table that offers definitions for each symbol. At times, the symbols have units or need units attached to them in order to explain the measure of a particular property. The International System of Units (SI) is usually used for science. Some of the most frequently used standard measures include: the meter (m) for length, the kilogram (kg) for mass, the ampere (A) for electric current, the kelvin (K) for temperature, and the mole (mol) for the amount of a substance. It never hurts to familiarize oneself with these commonly repeated symbols in scientific texts.

<u>Using Scientific Words to Express Science Information</u>

Numbers and symbols alone, however, do not fully convey scientific knowledge. Scientists rely heavily on specialized words to express scientific information. Presenting graphs, tables, and charts without any words would create a lot of confusion. Scientists use words to enhance the numbers and symbols they record and evaluate. Scientific literacy, therefore, requires a specialized synthesis of numbers, symbols, and scientific words. Many of these words are not used much outside the field of science.

Take the Linnaean system of classification as an example. It was created by Carl Linnaeus, a Swedish scientist working in the 18th century. This classification system helps scientists organize all living organisms into different phyla, classes, orders, families, genera, and species. The Linnaean scheme for organizing living organisms is still in use today. However, the entire scheme is founded upon a highly specialized naming system that draws on an archaic Latin language. The American coyote for example carries a common name—most people refer to it simply as the "coyote," whether they are using French, Spanish, or English. However, its taxonomic name in the Linnaean system is *Canis latrans*. A difficult name like *Canis latrans* illustrates just how specialized scientific words can be. For someone taking the exam, it is not necessary to memorize the specialized names of all species and all chemical compounds. However, it is certainly important to understand that this highly specialized language may emerge within a question to express scientific information. Together, along with symbols and numbers, these specialized words help create an entire scientific language. Scientific literacy at this level entails comprehending that such complex terms exist, and using context clues, dictionary definition keys, and other literacy-based tools to understand their meaning within the broader scope of the passage and its related assessment items.

Design of a Scientific Investigation

Science Investigations

<u>Designing a Science Investigation</u>

Human beings are, by nature, very curious. Long before the scientific method was established, people have been making observations and predicting outcomes, manipulating the physical world to create extraordinary things—from the first man-made fire in 6000 B.C.E. to the satellite that orbited Pluto in 2016. Although the history of the scientific method is sporadic and attributed to many different people, it remains the most reliable way to obtain and utilize knowledge about the observable universe. Designing a science investigation is based on the scientific method, which consists of the following steps:

- Make an observation
- Create a question
- Form a hypothesis
- Conduct an experiment
- Collect and analyze data
- Form a conclusion

The first step is to identify a problem based on an observation—the who, what, when, where, why, and how. An **observation** is the analysis of information using basic human senses: sight, sound, touch, taste, and smell. Observations can be two different types—qualitative or quantitative. A **qualitative observation** describes what is being observed, such as the color of a house or the smell of a flower. **Quantitative observations** measure what is being observed, such as the number of windows on a house or the intensity of a flower's smell on a scale of 1–5.

Observations lead to the identification of a problem, also called an **inference**. For example, if a fire truck is barreling down a busy street, the inferences could be:

- There's a fire.
- Someone is hurt.
- Some kid pulled the fire alarm at a local school.

Inferences are logical predictions based on experience or education that lead to the formation of a hypothesis.

Forming and Testing a Hypothesis

A hypothesis is a testable explanation of an observed scenario and is presented in the form of a statement. It's an attempt to answer a question based on an observation, and it allows a scientist to predict an outcome. A hypothesis makes assumptions on the relationship between two different variables, and answers the question: "If I do this, what happens to that?"

In order to form a hypothesis, there must be an independent variable and a dependent variable that can be measured. The **independent variable** is the variable that is manipulated, and the **dependent variable** is the result of the change.

For example, suppose a student wants to know how light affects plant growth. Based upon what he or she already knows, the student proposes (hypothesizes) that the more light to which a plant is exposed, the faster it will grow.

- Observation: Plants exposed to lots of light seem to grow taller.
- Question: Will plants grow faster if there's more light available?
- Hypothesis: The more light the plant has, the faster it will grow.
- Independent variable: The amount of time exposed to light (able to be manipulated)
- Dependent variable: Plant growth (the result of the manipulation)

Once a hypothesis has been formed, it must be tested to determine whether it's true or false. (How to test a hypothesis is described in a subsequent section.) After it has been tested and validated as true over and over, then a hypothesis can develop into a theory, model, or law.

Experimental Design

To test a hypothesis, one must conduct a carefully designed experiment. There are four basic requirements that must be present for an experiment to be valid:

- A control
- Variables
- A constant
- Repeated and collected data

The **control** is a standard to which the resultant findings are compared. It's the baseline measurement that allows for scientists to determine whether the results are positive or negative. For the example of light affecting plant growth, the control may be a plant that receives no light at all.

The independent variable is manipulated (a good way to remember this is *I* manipulate the *I*ndependent variable), and the dependent variable is the result of changes to the independent variable. In the plant example, the independent variable is the amount of time exposed to light, and the dependent variable is the resulting growth (or lack thereof) of the plant. For this experiment, there may be three plants—one that receives a minimal amount of light, the control, and one that receives a lot of light.

Finally, there must be constants in an experiment. A **constant** is an element of the experiment that remains unchanged. Constants are extremely important in minimizing inconsistencies within the experiment that may lead to results outside the parameters of the hypothesis. For example, some constants in the above case are that all plants receive the same amount of water, all plants are potted in the same kind of soil, the species of the plant used in each condition is the same, and the plants are stored at the same temperature. If, for instance, the plants received different amounts of water as well as light, it would be impossible to tell whether the plants responded to changes in water or light.

Once the experiment begins, a disciplined scientist must always record the observations in meticulous detail, usually in a journal. A good journal includes dates, times, and exact values of both variables and constants. Upon reading this journal, a different scientist should be able to clearly understand the experiment and recreate it exactly. The journal includes all collected data, or any observed changes. In this case, the data is rates of plant growth, as well as any other phenomena that occurred as a result of the experiment. A well-designed experiment also includes repetition in order to get the most accurate possible readings and to account for any errors, so several trials may be conducted.

Even in the presence of diligent constants, there are an infinite number of reasons that an experiment can (and will) go wrong, known as sources of error. All experimental results are inherently accepted as imperfect, if ever so slightly, because experiments are conducted by human beings, and no instrument can measure anything perfectly. The goal of scientists is to minimize those errors to the best of their ability.

<u>Identifying and Improving Hypotheses for Science Investigations</u>

When presented with fundamental, scientific concepts, it is important to read for understanding. The most basic skill in achieving this literacy is to understand the concept of hypothesis and moreover, to be able to identify it in a particular passage. A **hypothesis** is a proposed idea that needs further investigation in order to be proven true or false. While it can be considered an educated guess, a hypothesis goes more in depth in its attempt to explain something that is not currently accepted within scientific theory. It requires further experimentation and data gathering to test its validity and is subject to change, based on scientifically conducted test results. Being able to read a science passage and understand its main purpose, including any hypotheses, helps the test taker understand data-driven evidence. It helps the test taker to be able to correctly answer questions about the science excerpt they are asked to read.

When reading to identify a hypothesis, a test taker should ask, "What is the passage trying to establish? What is the passage's main idea? What evidence does the passage contain that either supports or refutes this idea?" Asking oneself these questions will help identify a hypothesis. Additionally, hypotheses are logical statements that are testable, and use very precise language.

Review the following example of a hypothesis:

> Consuming excess sugar in the form of beverages has a greater impact on childhood obesity and subsequent weight gain than excessive sugar from food.

While this is likely a true statement, it is still only a conceptual idea in a text passage regarding sugar consumption in childhood obesity, unless the passage also contains tested data that either proves or disproves the statement. A test taker could expect the rest of the passage to cite data proving that children who drink empty calories and don't exercise will, in fact, be obese.

A hypothesis goes further in that, given its ability to be proven or disproven, it may result in further hypotheses that require extended research. For example, the hypothesis regarding sugar consumption in drinks, after undergoing rigorous testing, may lead scientists to state another hypothesis such as the following:

> Consuming excess sugar in the form of beverages as opposed to food items is a habit found in mostly sedentary children.

This new, working hypothesis further focuses not just on the source of an excess of calories, but tries an "educated guess" that empty caloric intake has a direct, subsequent impact on physical behavior.

The data-driven chart below is similar to an illustration a test taker might see in relation to the hypothesis on sugar consumption in children:

Behaviors of Healthy and Unhealthy Kids

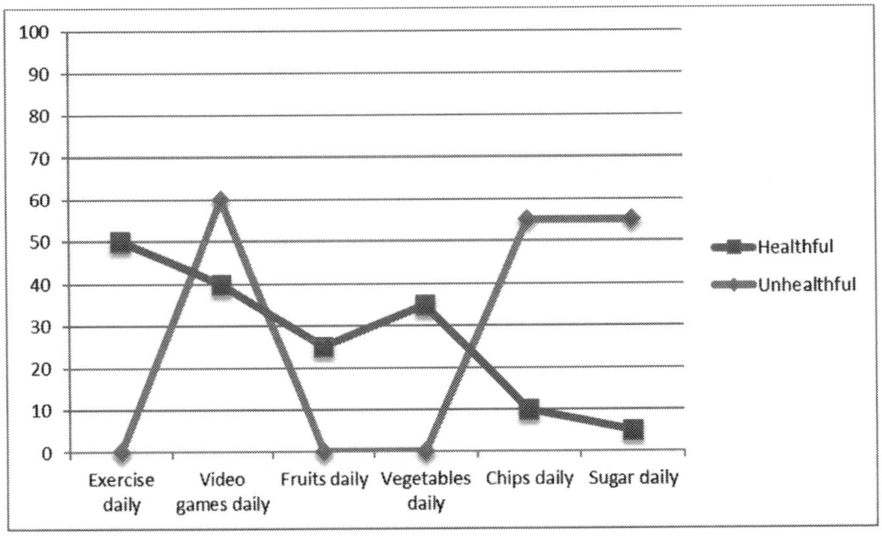

While this guide will address other data-driven passages a test taker could expect to see within a given science excerpt, note that the hypothesis regarding childhood sugar intake and rate of exercise has undergone scientific examination and yielded results that support its truth.

When reading a science passage to determine its hypothesis, a test taker should look for a concept that attempts to explain a phenomenon, is testable, logical, precisely worded, and yields data-driven results. The test taker should scan the presented passage for any word or data-driven clues that will help identify the hypothesis, and then be able to correctly answer test questions regarding the hypothesis based on their critical thinking skills.

Identifying Possible Errors in a Science Investigation and Changing the Design to Correct Them

For a hypothesis to be proven true or false, all experiments are subject to multiple trials in order to verify accuracy and precision. A measurement is **accurate** if the observed value is close to the "true value." For example, if someone measured the pH of water at 6.9, this measurement would be considered accurate (the pH of water is 7). On the other hand, a measurement is **precise** if the measurements are consistent—that is, if they are reproducible. If someone had a series of values for a pH of water that were 6.9, 7.0, 7.2, and 7.3, their measurements would not be precise. However, if all measured values were 6.9, or the average of these values was 6.9 with a small range, then their measurements would be precise.

Measurements can fall into the following categories:

- Both accurate and precise
- Accurate but not precise
- Precise but not accurate
- Neither accurate nor precise

The accuracy and precision of observed values most frequently correspond to the amount of error present in the experiment. Aside from general carelessness, there are two primary types of error:

random and systematic. **Random errors** are unpredictable variations in the experiment that occur by chance. They can be difficult to detect, but they can often be nullified using a statistical analysis and minimized by taking repeated measurements and taking an average. **Systematic errors** occur when there are imperfections in the design of the experiment itself—usually errors that affect the accuracy of the measurements. These errors can be minimized by using the most accurate equipment available and by taking proper care of instruments and measuring techniques. Common examples of error are listed below.

Random	Systematic
Environmental factors (random changes in vibration, temperature, humidity, etc.)	Poorly maintained instruments
	Old or out-of-date instruments
Differences in instrument use among scientists	Faulty calibration of instruments
Errors in judgment—can be affected by state of mind	Reading the instruments at an angle (parallax error) or other faulty reading errors
Incorrectly recorded observations	Not accounting for lag time

The most basic method to account for the possibility of errors is to take an average (also called a **mean**) of all observed values. To do so, one must divide the number of measurements taken from the sum of all measurements.

$$\frac{Sum\ of Measurements}{Total\ \#\ of\ Measurements}$$

For the above example of the pH values, the average is calculated by finding the sum of the pH values ascertained and dividing by the number of values recorded.

$$\frac{6.9 + 7.0 + 7.2 + 7.3}{4} = 7.1$$

The more observations recorded, the greater the precision. It's important to first assess the accuracy of measurements before proceeding to collect multiple trials of data. If a particular trial results in measurements that are vastly different from the average, it may indicate that a random or systematic error occurred during the trial. When this happens, a scientist might decide to "throw out" the trial and run the experiment again.

<u>Identifying the Strengths and Weaknesses of Different Types of Science Investigations</u>

In order to address the strengths and weaknesses of different types of scientific investigations, test takers must first strengthen their capacity for scientific literacy and numeracy. It is important to familiarize oneself with methods for decoding highly specialized scientific terms, formulas, and symbols. Additionally, test takers can take the following suggestions to help identify unique weaknesses and strengths in different types of scientific investigations:

- Using critical analysis, test takers begin asking questions about the accuracy of the methods used to collect, analyze, and display data. They should carefully look at text and graphics that show scientific findings.

- Test takers should determine whether or not the words, data, and symbols provided by the author actually offer information that is relevant for testing a hypothesis or making an inference.

- When two or more passages on the same topic are offered, test takers should cross-analyze the findings to determine what data is accurate or relevant and which findings are most objective.
- Although scientific research strives for objectivity, test takers should highlight any subjective biases that may be embedded in a text. In particular, they should be aware of certain historical or ethical biases that might appear.
- Test takers should double check for any computational inaccuracies.
- Test takers should make suggestions for better ways to present the findings in both texts and visual images.

Practice Test #1

1. Which statement about white blood cells is true?
   a. B cells are responsible for antibody production.
   b. White blood cells are made in the white/yellow cartilage before they enter the bloodstream.
   c. Platelets, a special class of white blood cell, function to clot blood and stop bleeding.
   d. The majority of white blood cells only activate during the age of puberty, which explains why children and the elderly are particularly susceptible to disease.

2. In testing how quickly a rat dies by the amount of poison it eats, which of the following is the independent variable and which is the dependent variable?
   a. How quickly the rat dies is the independent variable; the amount of poison is the dependent variable.
   b. The amount of poison is the independent variable; how quickly the rat dies is the dependent variable.
   c. Whether the rat eats the poison is the independent variable; how quickly the rat dies is the dependent variable.
   d. The cage the rat is kept in is the independent variable; the amount of poison is the dependent variable.

3. Which of the following are functions of the urinary system?
I. Synthesizing calcitriol and secreting erythropoietin
II. Regulating the concentrations of sodium, potassium, chloride, calcium, and other ions
III. Reabsorbing or secreting hydrogen ions and bicarbonate
IV. Detecting reductions in blood volume and pressure
   a. I, II, and III
   b. II and III
   c. II, III, and IV
   d. All of the above

4. Which of the following structures is unique to eukaryotic cells?
   a. Cell walls
   b. Nucleuses
   c. Cell membranes
   d. Vacuoles

5. Which is the cellular organelle used for digestion to recycle materials?
   a. The Golgi apparatus
   b. The lysosome
   c. The centrioles
   d. The mitochondria

6. A rock has a mass of 14.3 grams (g) and a volume of 5.4 cm^3, what is its density?
   a. 8.90 g/cm^3
   b. 0.38 g/cm^3
   c. 77.22 g/cm^3
   d. 2.65 g/cm^3

7. Why do arteries have valves?
   a. They have valves to maintain high blood pressure so that capillaries diffuse nutrients properly.
   b. Their valves are designed to prevent backflow due to their low blood pressure.
   c. They have valves due to a leftover trait from evolution that, like the appendix, are useless.
   d. They do not have valves, but veins do.

8. If the pressure in the pulmonary artery is increased above normal, which chamber of the heart will be affected first?
   a. The right atrium
   b. The left atrium
   c. The right ventricle
   d. The left ventricle

9. What is the purpose of sodium bicarbonate when released into the lumen of the small intestine?
   a. It works to chemically digest fats in the chyme.
   b. It decreases the pH of the chyme so as to prevent harm to the intestine.
   c. It works to chemically digest proteins in the chyme.
   d. It increases the pH of the chyme so as to prevent harm to the intestine.

10. Which of the following describes a reflex arc?
   a. The storage and recall of memory
   b. The maintenance of visual and auditory acuity
   c. The autoregulation of heart rate and blood pressure
   d. A stimulus and response controlled by the spinal cord

11. Describe the synthesis of the lagging strand of DNA.
   a. DNA polymerases synthesize DNA continuously after initially attaching to a primase.
   b. DNA polymerases synthesize DNA discontinuously in pieces called Okazaki fragments after initially attaching to primases.
   c. DNA polymerases synthesize DNA discontinuously in pieces called Okazaki fragments after initially attaching to RNA primers.
   d. DNA polymerases synthesize DNA discontinuously in pieces called Okazaki fragments which are joined together in the end by a DNA helicase.

12. Using anatomical terms, what is the relationship of the sternum relative to the deltoid?
   a. Medial
   b. Lateral
   c. Superficial
   d. Posterior

13. Ligaments connect what?
   a. Muscle to muscle
   b. Bone to bone
   c. Bone to muscle
   d. Muscle to tendon

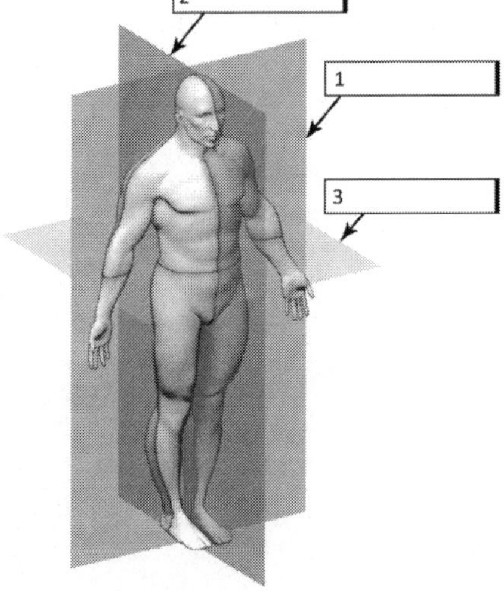

14. Identify the correct sequence of the 3 primary body planes as numbered 1, 2, and 3 in the above image.
    a. Plane 1 is coronal, plane 2 is sagittal, and plane 3 is transverse.
    b. Plane 1 is sagittal, plane 2 is coronal, and plane 3 is medial.
    c. Plane 1 is coronal, plane 2 is sagittal, and plane 3 is medial.
    d. Plane 1 is sagittal, plane 2 is coronal, and plane 3 is transverse.

15. Which of the following is NOT a major function of the respiratory system in humans?
    a. It provides a large surface area for gas exchange of oxygen and carbon dioxide.
    b. It helps regulate the blood's pH.
    c. It helps cushion the heart against jarring motions.
    d. It is responsible for vocalization.

16. Which of the following is NOT a function of the forebrain?
    a. To regulate blood pressure and heart rate
    b. To perceive and interpret emotional responses like fear and anger
    c. To perceive and interpret visual input from the eyes
    d. To integrate voluntary movement

17. What is the major difference between somatic and germline mutations?
    a. Somatic mutations usually benefit the individual while germline mutations usually harm them.
    b. Since germline mutations only affect one cell, they are less noticeable than the rapidly dividing somatic cells.
    c. Somatic mutations are not expressed for several generations, but germline mutations are expressed immediately.
    d. Germline mutations are usually inherited while somatic mutations will affect only the individual.

18. A child complains of heavy breathing even when relaxing. They are an otherwise healthy child with no history of respiratory problems. What might be the issue?
    a. Asthma
    b. Blood clot
    c. Hyperventilation
    d. Exercising too hard

19. Find the lowest coefficients that will balance the following combustion equation.

$$\_\_C_2H_{10}+\_\_O_2 \rightarrow \_\_H_2O+\_\_CO_2$$

a. 1:5:5:2
b. 4:10:20:8
c. 2:9:10:4
d. 2:5:10:4

20. What is the purpose of a catalyst?
a. To increase a reaction's rate by increasing the activation energy
b. To increase a reaction's rate by increasing the temperature
c. To increase a reaction's rate by decreasing the activation energy
d. To increase a reaction's rate by decreasing the temperature

21. Most catalysts found in biological systems are which of the following?
a. Special lipids called cofactors.
b. Special proteins called enzymes.
c. Special lipids called enzymes.
d. Special proteins called cofactors.

22. Which statement is true about the pH of a solution?
a. A solution cannot have a pH less than 1.
b. The more hydroxide ions in the solution, the higher the pH.
c. If an acid has a pH of greater than 2, it is considered a weak base.
d. A solution with a pH of 2 has ten times the amount of hydrogen ions than a solution with a power of 1.

23. Salts like sodium iodide (NaI) and potassium chloride (KCl) use what type of bond?
a. Ionic bonds
b. Disulfide bridges
c. Covalent bonds
d. London dispersion forces

24. Which of the following is unique to covalent bonds?
a. Most covalent bonds are formed between the elements H, F, N, and O.
b. Covalent bonds are dependent on forming dipoles.
c. Bonding electrons are shared between two or more atoms.
d. Molecules with covalent bonds tend to have a crystalline solid structure.

25. Which of the following describes a typical gas?
a. Indefinite shape and indefinite volume
b. Indefinite shape and definite volume
c. Definite shape and definite volume
d. Definite shape and indefinite volume

26. Which of the following areas of the body has the most sweat glands?
   a. Upper back
   b. Arms
   c. Feet
   d. Palms

27. A patient's body is not properly filtering blood. Which of the following body parts is most likely malfunctioning?
   a. Medulla
   b. Heart
   c. Nephrons
   d. Renal cortex

28. A pediatrician notes that an infant's cartilage is disappearing and being replaced by bone. What process has the doctor observed?
   a. Mineralization
   b. Ossification
   c. Osteoporosis
   d. Calcification

29. The epidermis is composed of what type of cells?
   a. Osteoclasts
   b. Connective
   c. Dendritic
   d. Epithelial

30. Which of the following is directly transcribed from DNA and represents the first step in protein building?
   a. siRNA
   b. rRNA
   c. mRNA
   d. tRNA

31. What information does a genotype give that a phenotype does not?
   a. The genotype necessarily includes the proteins coded for by its alleles.
   b. The genotype will always show an organism's recessive alleles.
   c. The genotype must include the organism's physical characteristics.
   d. The genotype shows what an organism's parents looked like.

	T	t
T		
t		

32. Which statement is supported by the Punnett square above, if "T" = Tall and "t" = short?
    a. Both parents are homozygous tall.
    b. 100% of the offspring will be tall because both parents are tall.
    c. There is a 25% chance that an offspring will be short.
    d. The short allele will soon die out.

33. Which of the following is a chief difference between evaporation and boiling?
    a. Liquids boil only at the surface while they evaporate equally throughout the liquid.
    b. Evaporating substances change from gas to liquid while boiling substances change from liquid to gas.
    c. Evaporation happens in nature while boiling is a manmade phenomenon.
    d. Evaporation can happen below a liquid's boiling point.

34. Which of the following CANNOT be found in a human cell's genes?
    a. Sequences of amino acids to be transcribed into mRNA
    b. Lethal recessive traits like sickle cell anemia
    c. Mutated DNA
    d. DNA that codes for proteins the cell doesn't use

35. Which of the following is a special property of water?
    a. Water easily flows through phospholipid bilayers.
    b. A water molecule's oxygen atom allows fish to breathe.
    c. Water is highly cohesive which explains its high melting point.
    d. Water can self-hydrolyze and decompose into hydrogen and oxygen.

36. What is an isotope?
    a. For any given element, it is an atom with a different atomic number.
    b. For any given element, it is an atom with a different number of protons.
    c. For any given element, it is an atom with a different number of electrons.
    d. For any given element, it is an atom with a different mass number.

37. What is the electrical charge of the nucleus?
    a. A nucleus always has a positive charge.
    b. A stable nucleus has a positive charge, but a radioactive nucleus may have no charge and instead be neutral.
    c. A nucleus always has no charge and is instead neutral.
    d. A stable nucleus has no charge and is instead neutral, but a radioactive nucleus may have a charge.

38. A student believes that there is an inverse relationship between sugar consumption and test scores. To test this hypothesis, he recruits several people to eat sugar, wait one hour, and take a short aptitude test afterwards. The student will compile the participants' sugar intake levels and test scores. How should the student conduct the experiment?

a. One round of testing, where each participant consumes a different level of sugar.

b. Two rounds of testing: The first, where each participant consumes a different level of sugar, and the second, where each participant consumes the same level as they did in Round 1.

c. Two rounds of testing: The first, where each participant consumes the same level of sugar as each other, and the second, where each participant consumes the same level of sugar as each other but at higher levels than in Round 1.

d. One round of testing, where each participant consumes the same level of sugar.

39. Which of the following creates sperm?

a. Prostate gland

b. Seminal vesicles

c. Scrotum

d. Seminiferous tubules

40. A researcher is exploring factors that contribute to the GPA of college students. While the sample is small, the researcher is trying to determine what the data shows. What can be reasoned from the table below?

Student	Maintains a Calendar?	Takes Notes?	GPA
A	sometimes	often	3.1
B	never	always	3.9
C	never	never	2.0
D	sometimes	often	2.7

a. No college students consistently maintain a calendar of events.

b. There is an inverse correlation between maintaining a calendar and GPA, and there is a positive correlation between taking notes and GPA.

c. There is a positive correlation between maintaining a calendar and GPA, and there is no correlation between taking notes and GPA.

d. There is no correlation between maintaining a calendar and GPA, and there is a positive correlation between taking notes and GPA.

41. Four different groups of the same species of peas are grown and exposed to differing levels of sunlight, water, and fertilizer as documented in the table below. The data in the water and fertilizer columns indicates how many times the peas are watered or fertilized per week, respectively. Group 2 is the only group that withered. What is a reasonable explanation for this occurrence?

Group	**Sunlight**	**Water**	**Fertilizer**
1	partial sun	4 mL/hr	1
2	full sun	7 mL/hr	1
3	no sun	14 mL/hr	2
4	partial sun	3 mL/hr	2

a. Insects gnawed away the stem of the plant.
b. The roots rotted due to poor drainage.
c. The soil type had nutrition deficiencies.
d. This species of peas does not thrive in full sunlight.

42. Which of the following functions corresponds to the parasympathetic nervous system?
a. It stimulates the fight-or-flight response.
b. It increases heart rate.
c. It stimulates digestion.
d. It increases bronchiole dilation.

43. According to the periodic table, which of the following elements is the least reactive?
a. Fluorine
b. Silicon
c. Neon
d. Gallium

44. The Human Genome Project is a worldwide research project launched in 1990 to map the entire human genome. Although the Project was faced with the monumental challenge of analyzing tons and tons of data, its objective was completed in 2003 and ahead of its deadline by two years. Which of the following inventions likely had the greatest impact on this project?
a. The sonogram
b. X-ray diffraction
c. The microprocessor
d. Magnetic Resonance Imaging (MRI)

45. Which of the following inventions likely had the greatest improvement on the ability to combat nutrition deficiencies in developing countries?
a. Food products fortified with dietary vitamins and minerals
b. Integrated statistical models of fish populations
c. Advances so that microscopes can use thicker tissue samples
d. Refrigerated train cars for transportation of food

46. Which of the following is the gland that helps regulate calcium levels?
  a. Osteoid gland
  b. Pineal gland
  c. Parathyroid glands
  d. Thymus gland

47. Anya was paid by Company X to analyze dwindling honeybee populations of the Southwest. After measuring hive populations over several months, she noticed no patterns in the geographic distributions of the deaths after comparisons with local maps of interest. This supported her hypothesis, so she took samples of the honey and the bees from the hives and performed dozens of dissections to confirm her suspicions. Which of the following is the most likely hypothesis upon which this research was performed?
  a. Honeybees are being killed off and their hives destroyed by other extremely aggressive species of bees from the South.
  b. Honeybees are contracting parasites in large droves.
  c. Honeybees are so sensitive to certain pesticides that they die on contact.
  d. Honeybees die in larger numbers around cell phone towers.

48. Explain the Law of Conservation of Mass as it applies to this reaction: $2H_2 + O_2 \rightarrow 2H_2O$.
  a. Electrons are lost.
  b. The hydrogen loses mass.
  c. New oxygen atoms are formed.
  d. There is no decrease or increase of matter

49. Which of the following is the best unit to measure the amount of blood in the human body?
  a. Ounces
  b. Liters
  c. Milliliters
  d. Pounds

50. Which of the following systems does not include a transportation system throughout the body?
  a. Cardiovascular system
  b. Endocrine system
  c. Immune system
  d. Nervous system

51. Which of the following correctly identifies a difference between the primary and secondary immune response?
  a. In the secondary response, macrophages migrate to the lymph nodes to present the foreign microorganism to helped T lymphocytes.
  b. The humeral immunity that characterizes the primary response is coordinated by T lymphocytes.
  c. The primary response is quicker and more powerful than the secondary response.
  d. Suppressor T cells activate in the secondary response to prevent an overactive immune response.

52. Eosinophils are best described as which of the following?
   a. A type of granulocyte that secretes histamine, which stimulates the inflammatory response.
   b. The most abundant type of white blood cell and they secrete substances that are toxic to pathogens.
   c. A type of granulocyte found under mucous membranes and defends against multicellular parasites.
   d. A type of circulating granulocyte that is aggressive and has high phagocytic activity.

53. Which of the following correctly matches a category of protein with a physiologic example?
   a. Keratin is a structural protein
   b. Antigens are hormonal proteins
   c. Channel proteins are marker proteins
   d. Actin is a transport protein

Answer Explanations #1

1. A: When activated, B cells create antibodies against specific antigens. White blood cells are generated in yellow bone marrow, not cartilage. Platelets are not a type of white blood cell and are typically cell fragments produced by megakaryocytes. White blood cells are active throughout nearly all of one's life, and have not been shown to specially activate or deactivate because of life events like puberty or menopause.

2. B: The independent variable is the variable manipulated and the dependent variable is the result of the changes in the independent variable. Choice *B* is correct because the amount of poison is the variable that is changed, and the speed of rat death is the result of the changes in the amount of poison administered. Choice *A* is incorrect because that answer states the opposite. Choice *C* is false because the scientist isn't attempting to determine whether the rat will die *if* it eats poison; the scientist is testing how quickly the rat will die depending on *how much* poison it eats. Choice *D* is incorrect because the cage isn't manipulated in any way and has nothing to do with the hypothesis.

3. D: The urinary system has many functions, the primary of which is removing waste products and balancing water and electrolyte concentrations in the blood. It also plays a key role in regulating ion concentrations, such as sodium, potassium, chloride, and calcium, in the filtrate. The urinary system helps maintain blood pH by reabsorbing or secreting hydrogen ions and bicarbonate as necessary. Certain kidney cells can detect reductions in blood volume and pressure and then can secrete renin to activate a hormone that causes increased reabsorption of sodium ions and water. This serves to raise blood volume and pressure. Kidney cells secrete erythropoietin under hypoxic conditions to stimulate red blood cell production. They also synthesize calcitriol, a hormone derivative of vitamin D3, which aids in calcium ion absorption by the intestinal epithelium.

4. B: The structure exclusively found in eukaryotic cells is the nucleus. Animal, plant, fungi, and protist cells are all eukaryotic. DNA is contained within the nucleus of eukaryotic cells, and they also have membrane-bound organelles that perform complex intracellular metabolic activities. Prokaryotic cells (archaea and bacteria) do not have a nucleus or other membrane-bound organelles and are less complex than eukaryotic cells.

5. B: The cell structure responsible for cellular storage, digestion and waste removal is the lysosome. Lysosomes are like recycle bins. They are filled with digestive enzymes that facilitate catabolic reactions to regenerate monomers. The Golgi apparatus is designed to tag, package, and ship out proteins destined for other cells or locations. The centrioles typically play a large role only in cell division when they ratchet the chromosomes from the mitotic plate to the poles of the cell. The mitochondria are involved in energy production and are the powerhouses of the cell.

6. D: Density is found by dividing mass by volume:

$$density = \frac{mass}{volume}$$

The unit for mass (in this case grams) and the units for volume (in this case cm^3) need to be combined together. They're combined as grams over volume since that's how they were set up in the equation:

$$d = \frac{14.3\ g}{5.4\ cm^3} = 2.65\ g/cm^3$$

7. D: Veins have valves, but arteries do not. Valves in veins are designed to prevent backflow, since they are the furthest blood vessels from the pumping action of the heart and steadily increase in volume (which decreases the available pressure). Capillaries diffuse nutrients properly because of their thin walls and high surface area and are not particularly dependent on positive pressure.

8. C: The blood leaves the right ventricle through a semi-lunar valve and goes through the pulmonary artery to the lungs. Any increase in pressure in the artery will eventually affect the contractibility of the right ventricle. Blood enters the right atrium from the superior and inferior venae cava veins, and blood leaves the right atrium through the tricuspid valve to the right ventricle. Blood enters the left atrium from the pulmonary veins carrying oxygenated blood from the lungs. Blood flows from the left atrium to the left ventricle through the mitral valve and leaves the left ventricle through a semi-lunar valve to enter the aorta.

9. D: Sodium bicarbonate, a very effective base, has the chief function to increase the pH of the chyme. Chyme leaving the stomach has a very low pH, due to the high amounts of acid that are used to digest and break down food. If this is not neutralized, the walls of the small intestine will be damaged and may form ulcers. Sodium bicarb is produced by the pancreas and released in response to pyloric stimulation so that it can neutralize the acid. It has little to no digestive effect.

10. D: A reflex arc is a simple nerve pathway involving a stimulus, a synapse, and a response that is controlled by the spinal cord—not the brain. The knee-jerk reflex is an example of a reflex arc. The stimulus is the hammer touching the tendon, reaching the synapse in the spinal cord by an afferent pathway. The response is the resulting muscle contraction reaching the muscle by an efferent pathway. None of the remaining processes is a simple reflex. Memories are processed and stored in the hippocampus in the limbic system. The visual center is located in the occipital lobe, while auditory processing occurs in the temporal lobe. The sympathetic and parasympathetic divisions of the autonomic nervous system control heart and blood pressure.

11. C: The lagging strand of DNA falls behind the leading strand because of its discontinuous synthesis. DNA helicase unzips the DNA helices so that synthesis can take place, and RNA primers are created by the RNA primase for the polymerases to attach to and build from. The lagging strand is synthesizing DNA in a direction that is hard for the polymerase to build, so multiple primers are laid down so that the entire length of DNA can be synthesized simultaneously, piecemeal. These short pieces of DNA being synthesized are known as Okazaki fragments and are joined together by DNA ligase.

12. A: The sternum is medial to the deltoid because it is much closer (typically right on) the midline of the body, while the deltoid is lateral at the shoulder cap. Superficial means that a structure is closer to the body surface and posterior means that it falls behind something else. For example, skin is superficial to bone and the kidneys are posterior to the rectus abdominus.

13. B: Ligaments connect bone to bone. Tendons connect muscle to bone. Both are made of dense, fibrous connective tissue (primary Type 1 collagen) to give strength. However, tendons are more organized, especially in the long axis direction like muscle fibers themselves, and they have more collagen. This arrangement makes more sense because muscles have specific orientations of their fibers, so they contract in somewhat predictable directions. Ligaments are less organized and more of a woven pattern because bone connections are not as organized as bundles or muscle fibers, so ligaments must have strength in multiple directions to protect against injury.

14. A: The three primary body planes are coronal, sagittal, and transverse. The coronal or frontal plane, named for the plane in which a corona or halo might appear in old paintings, divides the body vertically into front and back sections. The sagittal plane, named for the path an arrow might take when shot at the body, divides the body vertically into right and left sections. The transverse plane divides the body horizontally into upper or superior and lower or inferior sections. There is no medial plane, per se. The anatomical direction medial simply references a location close or closer to the center of the body than another location.

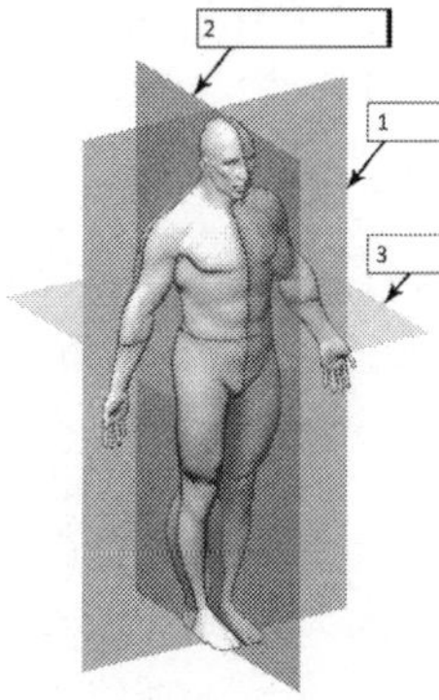

15. C: Although the lungs may provide some cushioning for the heart when the body is violently struck, this is not a major function of the respiratory system. Its most notable function is that of gas exchange for oxygen and carbon dioxide, but it also plays a vital role in the regulation of blood pH. The aqueous form of carbon dioxide, carbonic acid, is a major pH buffer of the blood, and the respiratory system directly controls how much carbon dioxide stays and is released from the blood through respiration. The respiratory system also enables vocalization and forms the basis for the mode of speech and language used by most humans.

16. A: The forebrain contains the cerebrum, the thalamus, the hypothalamus, and the limbic system. The limbic system is chiefly responsible for the perception of emotions through the amygdale, while the cerebrum interprets sensory input and generates movement. Specifically, the occipital lobe receives visual input, and the primary motor cortex in the frontal lobe is the controller of voluntary movement. The hindbrain, specifically the medulla oblongata and brain stem, control and regulate blood pressure and heart rate.

17. D: Germline mutations in eggs and sperm are permanent, can be on the chromosomal level, and will be inherited by offspring. Somatic mutations cannot affect eggs and sperm, and therefore are not inherited by offspring. Mutations of either kind are rarely beneficial to the individual, but do not necessarily harm them. Germline cells divide much more rapidly than do somatic cells, and a mutation in a sex cell would promulgate and affect many thousands of its daughter cells.

18. A: It is most likely asthma. Any of the answer choices listed can cause heavy breathing. A blood clot in the lung (*B*) could cause this, but this would be very uncommon for a child. Choices *C* and *D* can both be ruled out because the question mentions that it occurs even when the patient is relaxing. Hyperventilation is usually caused by a panic attack or some sort of physical activity. Asthma often develops during childhood. It would stand to reason then that the child may have not yet been diagnosed. While asthma attacks can be caused by exercise they can also occur when a person is not exerting themselves.

19. C: 2:9:10:4. These are the coefficients that follow the law of conservation of matter. The coefficient times the subscript of each element should be the same on both sides of the equation.

20. C: A catalyst functions to increase reaction rates by decreasing the activation energy required for a reaction to take place. Inhibitors would increase the activation energy or otherwise stop the reactants from reacting. Although increasing the temperature usually increases a reaction's rate, this is not true in all cases, and most catalysts do not function in this manner.

21. B: Biological catalysts are termed *enzymes*, which are proteins with conformations that specifically manipulate reactants into positions which decrease the reaction's activation energy. Lipids do not usually affect reactions, and cofactors, while they can aid or be necessary to the proper functioning of enzymes, do not make up the majority of biological catalysts.

22. B: Substances with higher amounts of hydrogen ions will have lower pHs, while substances with higher amounts of hydroxide ions will have higher pHs. Choice *A* is incorrect because it is possible to have an extremely strong acid with a pH less than 1, as long as its molarity of hydrogen ions is greater than 1. Choice *C* is false because a weak base is determined by having a pH lower than some value, not higher. Substances with pHs greater than 2 include anything from neutral water to extremely caustic lye. Choice *D* is false because a solution with a pH of 2 has ten times fewer hydrogen ions than a solution of pH 1.

23. A: Salts are formed from compounds that use ionic bonds. Disulfide bridges are special bonds in protein synthesis which hold the protein in their secondary and tertiary structures. Covalent bonds are strong bonds formed through the sharing of electrons between atoms and are typically found in organic molecules like carbohydrates and lipids. London dispersion forces are fleeting, momentary bonds which occur between atoms that have instantaneous dipoles but quickly disintegrate.

24. C: As in the last question, covalent bonds are special because they share electrons between multiple atoms. Most covalent bonds are formed between the elements H, F, N, O, S, and C, while hydrogen bonds are formed nearly exclusively between H and either O, N, or F. Covalent bonds may inadvertently form dipoles, but this does not necessarily happen. With similarly electronegative atoms like carbon and hydrogen, dipoles do not form, for instance. Crystal solids are typically formed by substances with ionic bonds like the salts sodium iodide and potassium chloride.

25. A: Gases like air will move and expand to fill their container, so they are considered to have an indefinite shape and indefinite volume. Liquids like water will move and flow freely, so their shapes change constantly, but do not change volume or density on their own. Solids change neither shape nor volume without external forces acting on them, so they have definite shapes and volumes.

26. A: The upper back has the one of the high densities of sweat glands of any area on the body. While palms, arms, and feet are often thought of as sweaty areas, they have relatively low amounts of sweat glands compared to other parts of the body. Remember that one of the purposes of sweat is thermoregulation, or controlling the temperature of the body. Regulating the temperature of one's core is more important than adjusting the temperature of one's extremities.

27. C: Nephrons are responsible for filtering blood. When functioning properly they allow blood cells and nutrients to go back into the bloodstream while sending waste to the bladder. However, nephrons can fail at doing this, particularly when blood flood to the kidneys is limited. The medulla (also called the renal medulla) (*A*) and the renal cortex (*D*) are both parts of the kidney but are not specifically responsible for filtering blood. The medulla is in the inner part of the kidney and contains the nephrons. The renal cortex is the outer part of the kidney. The heart (*B*) is responsible for pumping blood throughout the body rather than filtering it.

28. B: Ossification is the process by which cartilage, a soft, flexible substance is replaced by bone throughout the body. All humans regardless of age have cartilage, but cartilage in some areas goes away to make way for bones.

29. D: The outermost layer of the skin, the: epidermis, consists of epithelial cells. This layer of skin is dead, as it has no blood vessels. Osteoclasts are cells that make up bones. Notice the prefix *Osteo-* which means bone. Connective tissue macrophage cells can be found in a variety of places, and dendritic cells are part of the lymphatic system.

30. C: mRNA is directly transcribed from DNA before being taken to the cytoplasm and translated by rRNA into a protein. tRNA transfers amino acids from the cytoplasm to the rRNA for use in building these proteins. siRNA is a special type of RNA which interferes with other strands of mRNA typically by causing them to get degraded by the cell rather than translated into protein.

31. B: Since the genotype is a depiction of the specific alleles that an organism's genes code for, it includes recessive genes that may or may not be otherwise expressed. The genotype does not have to name the proteins that its alleles code for; indeed, some of them may be unknown. The phenotype is the physical, visual manifestations of a gene, not the genotype. The genotype does not necessarily include any information about the organism's physical characters. Although some information about an organism's parents can be obtained from its genotype, its genotype does not actually show the parents' phenotypes.

32. C: One in four offspring (or 25%) will be short, so all four offspring cannot be tall. Although both of the parents are tall, they are hybrid or heterozygous tall, not homozygous. The mother's phenotype is for tall, not short. A Punnett square cannot determine if a short allele will die out. Although it may seem intuitive that the short allele will be expressed by lower numbers of the population than the tall allele, it still appears in 75% of the offspring (although its effects are masked in 2/3 of those). Besides, conditions could favor the recessive allele and kill off the tall offspring.

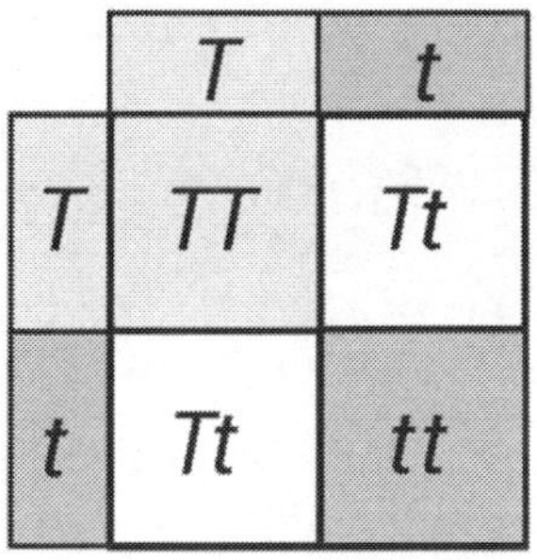

Both parents are mixed hybrids
so offspring are a 3:1 ratio

33. D: Evaporation takes place at the surface of a fluid while boiling takes place throughout the fluid. The liquid will boil when it reaches its boiling or vaporization temperature, but evaporation can happen due to a liquid's volatility. Volatile substances often coexist as a liquid and as a gas, depending on the pressure forced on them. The phase change from gas to liquid is condensation, and both evaporation and boiling take place in nature.

34. A: Human genes are strictly DNA and do not include proteins or amino acids. A human's genome and collection of genes will include even their recessive traits, mutations, and unused DNA.

35. C: Water's polarity lends it to be extremely cohesive and adhesive; this cohesion keeps its atoms very close together. Because of this, it takes a large amount of energy to melt and boil its solid and liquid forms. Phospholipid bilayers are made of nonpolar lipids and water, a polar liquid, cannot flow through it. Cell membranes use proteins called aquaporins to solve this issue and let water flow in and out. Fish breathe by capturing dissolved oxygen through their gills. Water can self-ionize, wherein it decomposes into a hydrogen ion (H^+) and a hydroxide ion (OH^-), but it cannot self-hydrolyze.

36. D: An isotope of an element has an atomic number equal to its number of protons, but a different mass number because of the additional neutrons. Even though there are differences in the nucleus, the behavior and properties of isotopes of a given element are identical. Atoms with different atomic numbers also have different numbers of protons and are different elements, so they cannot be isotopes.

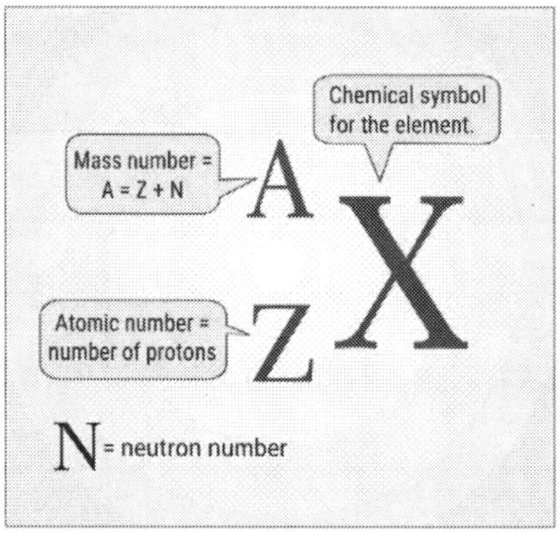

37. A: The neutrons and protons make up the nucleus of the atom. The nucleus is positively charged due to the presence of the protons. The negatively charged electrons are attracted to the positively charged nucleus by the electrostatic or Coulomb force; however, the electrons are not contained in the nucleus. The positively charged protons create the positive charge in the nucleus, and the neutrons are electrically neutral, so they have no effect. Radioactivity does not directly have a bearing on the charge of the nucleus.

38. C: To gather accurate data, the student must be able compare a participant's test score from round 1 with their test score from round 2. The differing levels of intellect among the participants means that comparing participants' test scores to those of other participants would be inaccurate. This requirement excludes choices A and D, which involve only one round of testing. The experiment must also involve different levels of sugar consumption from round 1 to round 2. In this way, the effects of different levels of sugar consumption can be seen on the same subjects. Thus, B is incorrect because the experiment provides for no variation of sugar consumption. C is the correct answer because it allows the student to compare each participant's test score from round 1 with their test score from round 2 after different level of sugar consumption.

39. D: The seminiferous tubules are responsible for sperm production. Had *testicles* been an answer choice, it would also have been correct since it houses the seminiferous tubules. The prostate gland (*A*) secretes enzymes that help nourish sperm after creation. The seminal vesicles (*B*) secrete some of the components of semen. The scrotum (*C)* is the pouch holding the testicles.

40. D: Based on this table, it can be reasoned that there is not a correlation between maintaining a calendar and GPA, since Student B never maintains a calendar but has the highest GPA of the cohort. Furthermore, it can be reasoned that there is a positive correlation between taking notes and GPA since the more notes a student takes, the higher the GPA they have Thus, Choice *D* is the correct answer.

Choice *A* offers an absolute that cannot be proven based on this study; thus, it is incorrect. Choices B and C are incorrect because they have at least one incorrect correlation.

41. D: *D* is the correct answer because excess sunlight is a common cause of plant wilting. *A*, *B*, and *C* are all possible but unlikely to be a cause for wilting. Given that the test question asks for a *reasonable* explanation, *sunlight* is by far the most reasonable answer.

42. C: The parasympathetic nervous system is related to calm, peaceful times without stress that require no immediate decisions. It relaxes the fight-or-flight response, slows heart rate to a comfortable pace, and decreases bronchiole dilation to a normal size. The sympathetic nervous system, on the other hand, is in charge of the fight-or-flight response and works to increase blood pressure and oxygen absorption.

43. C: Neon, one of the noble gases, is chemically inert or not reactive because it contains eight valence electrons in the outermost shell. The atomic number is 10, with a 2.8 electron arrangement meaning that there are 2 electrons in the inner shell and the remaining 8 electrons in the outer shell. This is extremely stable for the atom, so it will not want to add or subtract any of its electrons and will not react under typical circumstances.

44. C: Because of the vast amounts of data that needed to be processed and analyzed, technological breakthroughs like innovations to the microprocessor were directly responsible for the ease of computing handled by the Human Genome Project. Although the sonogram and MRI technology are helpful to the healthcare industry in general, they would not have provided a great deal of help for sequencing and comprehending DNA data, in general. X-ray diffraction is a technique that helps visualize the structures of crystallized proteins, but cannot determine DNA bases with enough precision to help sequence DNA.

45. A: Many foods from developed countries are grown from plants which have been processed or bioengineered to include increased amounts of nutrients like vitamins and minerals that otherwise would be lost during manufacturing or are uncommon to the human diet. White rice, for example, is typically enriched with niacin, iron, and folic acid, while salt has been fortified with iodine for nearly a century. These help to prevent nutrition deficiencies. While it can be useful for fisheries to maintain models of fish populations so that they don't overfish their stock, this is not as immediately important to nutrition as are fortified and enriched foods. Although innovations to microscopes could lead to improved healthcare, this also has no direct effect on nutrition deficiency. Refrigerated train carts were historically a crucial invention around Civil War times and were used to transport meat and dairy products long distances without spoiling, but dietary deficiencies could be more easily remedied by supplying people with fortified foods containing those nutrients rather than spoilable meats.

46. C: The parathyroid gland impacts calcium levels by secreting parathyroid hormone (PTH). Osteoid gland is not a real gland. The pineal gland regulates sleep by secreting melatonin, and the thymus gland focuses on immunity. *Thyroid* would also be a correct answer choice as it influences the levels of circulating calcium.

47. B: The most likely hypothesis that Anya is testing has something to with the pathophysiology of the bees, as she performed dissections on some of her samples. These dissections would be unnecessary if the bees were being killed off by another species, as the destruction of the hives would be obvious. As the deaths did not seem linked in any particular way geographically, it is also safe to assume that there was no correlation to cell phone towers, as maps of cell phone coverage would be readily available to her. If the pesticides were so toxic that the bees died on contact, then they wouldn't make it back to the hive to be available for her dissections or to drop off chemicals in the honey. The most likely of the choices is that parasites are killing off the bees, which would be easily communicable.

48. D: The law states that matter cannot be created or destroyed in a closed system. In this equation, there are the same number of molecules of each element on either side of the equation. Matter is not gained or lost, although a new compound is formed. As there are no ions on either side of the equation, no electrons are lost. The law prevents the hydrogen from losing mass and prevents oxygen atoms from being spontaneously spawned.

49. B: For measuring blood, we're looking for a unit that measures volume. Choices *A, B*, and *C*, are all measures of volume, but pounds (*D*) is a measure of weight. The correct answer is liters, as the average adult has about 5 liters of blood in their body. Blood can certainly be measured in ounces or milliliters; however, 5 liters is equal to 5,000 milliliters or 176 ounces. Thus, liters seems to be the more rational measuring unit.

50. B: The endocrine system's organs are glands which are spread throughout the body. The endocrine system itself does not connect the organs or transport the hormones they secrete. Rather, the various glands secrete the hormone into the bloodstream and lets the cardiovascular system pump it throughout the body. The other three body systems each include a network throughout the body:

- Cardiovascular system: veins and arteries
- Immune system: lymphatic vessels (it does also use the circulatory system)
- Nervous system: nerve networks

51. D: In the secondary immune response, suppressor T lymphocytes are activated to negate the potential risk of damage to healthy cells, brought on by an unchecked, overactive immune response. Choice *A* is incorrect because the activity is characteristic of the primary response, not the secondary response Choice *B* is incorrect because humeral immunity is mediated by antibodies produced by B, not T, lymphocytes. Choice *C* is wrong because the secondary response is faster than the primary response because the primary response entails the time-consuming process of macrophage activation.

52. C: Eosinophils, like neutrophils, basophils, and mast cells, are a type of leukocyte in a class called granulocytes. They are found underneath mucous membranes in the body and they primarily secrete destructive enzymes and defend against multicellular parasites like worms. Choice *A* describes basophils and mast cells, and Choice *B* and *D* describe neutrophils. Unlike neutrophils which are aggressive phagocytic cells, eosinophils have low phagocytic activity.

53. A: Keratin is a structural protein and it is the primary constituent of things like hair and nails. Choice *B* is incorrect antigens are immune proteins that help fight disease. Hormonal proteins are responsible for initiating the signal transduction cascade to regulate gene expression. Choice *C* is incorrect because channel proteins are transport proteins that help move molecules into and out of a cell. Marker proteins help identify or distinguish a cell. Lastly, Choice *D* is incorrect because actin, like myosin, is a motor protein because it is involved in the process of muscle contraction.

Practice Test #2

1. A scientist is trying to determine how much poison will kill a rat the fastest. Which of the following statements is an example of an appropriate hypothesis?
   a. Rats that are given lots of poison seem to die quickly.
   b. Does the amount of poison affect how quickly the rat dies?
   c. The more poison a rat is given, the quicker it will die.
   d. Poison is fatal to rats.

2. Which locations in the digestive system are sites of chemical digestion?
   I. Mouth
   II. Stomach
   III. Small Intestine

   a. II only
   b. III only
   c. II and III only
   d. I, II, and III

3. Which of the following is a representation of a natural pattern or occurrence that's difficult or impossible to experience directly?
   a. A theory
   b. A model
   c. A law
   d. An observation

4. Which of the following is a standard or series of standards to which the results from an experiment are compared?
   a. A control
   b. A variable
   c. A constant
   d. Collected data

5. "This flower is dead; someone must have forgotten to water it." This statement is an example of which of the following?
   a. A classification
   b. An observation
   c. An inference
   d. A collection

6. Which of the following correctly displays 8,600,000,000,000 in scientific notation (to two significant figures)?
   a. 8.6×10^{12}
   b. 8.6×10^{-12}
   c. 8.6×10^{11}
   d. 8.60×10^{12}

7. The acceleration of a falling object due to gravity has been proven to be 9.8 m/s^2. A scientist drops a cactus four times and measures the acceleration with an accelerometer and gets the following results: 9.79 m/s^2, 9.81 m/s^2, 9.80 m/s^2, and 9.78 m/s^2. Which of the following accurately describes the measurements?

a. They're both accurate and precise.
b. They're accurate but not precise.
c. They're precise but not accurate.
d. They're neither accurate nor precise.

8. What is the molarity of a solution made by dissolving 4.0 grams of NaCl into enough water to make 120 mL of solution? The atomic mass of Na is 23.0 g/mol and Cl is 35.5 g/mol.

a. 0.34 M
b. 0.57 M
c. 0.034 M
d. 0.057 M

9. Considering a gas in a closed system, at a constant volume, what will happen to the temperature if the pressure is increased?

a. The temperature will stay the same
b. The temperature will decrease
c. The temperature will increase
d. It cannot be determined with the information given

10. What type of chemical reaction produces a salt?

a. An oxidation reaction
b. A neutralization reaction
c. A synthesis reaction
d. A decomposition reaction

11. What types of molecules can move through a cell membrane by passive transport?

a. Complex sugars
b. Non-lipid soluble molecules
c. Oxygen
d. Molecules moving from areas of low concentration to areas of high concentration

12. What is ONE feature that both prokaryotes and eukaryotes have in common?

a. A plasma membrane
b. A nucleus enclosed by a membrane
c. Organelles
d. A nucleoid

13. What is the LAST phase of mitosis?

a. Prophase
b. Telophase
c. Anaphase
d. Metaphase

14. How many daughter cells are formed from one parent cell during meiosis?
   a. One
   b. Two
   c. Three
   d. Four

15. Which level of protein structure is defined by the folds and coils of the protein's polypeptide backbone?
   a. Primary
   b. Secondary
   c. Tertiary
   d. Quaternary

16. What is the sensory threshold?
   a. The smallest amount of stimulus required for an individual to feel a sensation
   b. The amount of stimulus required for an individual to feel pain
   c. The amount of stimulus required to cause an individual to move away from the stimulus
   d. The place where the stimulus is coming from

17. How many neurons generally make up a sensory pathway?
   a. 1
   b. 2
   c. 3
   d. 4

18. In which part of the eye does visual processing begin?
   a. Cornea
   b. Optic nerve
   c. Retina
   d. Eyelid

19. What type of vessel carries oxygen-rich blood from the heart to other tissues of the body?
   a. Veins
   b. Intestines
   c. Bronchioles
   d. Arteries

20. The somatic nervous system is responsible for which of the following?
   a. Breathing
   b. Thought
   c. Movement
   d. Fear

21. The process of breaking large molecules into smaller molecules to provide energy is known as which of the following?
   a. Metabolism
   b. Bioenergetics
   c. Anabolism
   d. Catabolism

22. Which blood component is chiefly responsible for clotting?
    a. Platelets
    b. Red blood cells
    c. Antigens
    d. Plasma cells

23. Which is the first event to happen in a primary immune response?
    a. Macrophages phagocytose pathogens and present their antigens.
    b. Neutrophils aggregate and act as cytotoxic, nonspecific killers of pathogens.
    c. B lymphocytes make pathogen-specific antibodies.
    d. Helper T cells secrete interleukins to activate pathogen-fighting cells.

24. Where does sperm maturation take place in the male reproductive system?
    a. Seminal vesicles
    b. Prostate gland
    c. Epididymis
    d. Vas Deferens

25. Which of the following lists of joint types is in the correct order for increasing amounts of permitted motion (least mobile to most mobile)?
    a. Hinge, condyloid, saddle
    b. Saddle, hinge, condyloid
    c. Saddle, condyloid, hinge
    d. Hinge, saddle, condyloid

26. The function of synergists can best be described as which of the following?
    I. They assist primary movers in completing the specific movement
    II. They stabilize the point of origin and provide extra pull near the insertion
    III. They help prevent unwanted movement at a joint

    a. I, II
    b. I, III
    c. II, III
    d. All of the above

27. What makes bone resistant to shattering?
    a. The calcium salts deposited in the bone
    b. The collagen fibers
    c. The bone marrow and network of blood vessels
    d. The intricate balance of minerals and collagen fibers

28. How do organisms maintain homeostasis?
    a. They increase their body temperature, blood pH, and fluid balance.
    b. They undergo biochemical processes and absorb energy to increase entropy.
    c. They undergo biochemical processes to maintain the order of their external environment.
    d. They use free energy and matter via biochemical processes to work against entropy.

29. Which of the following correctly lists the four properties that all types of muscle tissue share?
   a. Contractile, excitable, elastic, extensible
   b. Contractile, voluntary, elastic, extensible
   c. Contractile, excitable, voluntary, extensible
   d. Contractile, excitable, elastic, voluntary

30. Which of the following about the autonomic nervous system (ANS) is true?
   a. It controls the reflex arc
   b. It contains motor (efferent) neurons
   c. It contains sensory (afferent) neurons
   d. It contains both parasympathetic nerves and sympathetic nerves

31. What is the function of the sinuses?
   a. To trap the many airborne pathogens
   b. To direct air down the trachea rather than the esophagus
   c. To warm, humidify, and filter air
   d. To sweep away pathogens and direct them toward the top of the trachea

32. Which of the following structures acts like a funnel by delivering the urine from the millions of the collecting tubules to the ureters?
   a. The renal pelvis
   b. The renal cortex
   c. The renal medulla
   d. Bowman's capsule

33. Which of the following is the body cavity that contains the urinary bladder, urethra, and ureters?
   a. The thoracic cavity
   b. The pelvic cavity
   c. The abdominal cavity
   d. The spinal cavity

34. Which of the following best defines the term *amphoteric*?
   a. A substance that conducts electricity due to ionization when dissolved in a solvent
   b. A substance that can act as an acid or a base depending on the properties of the solute
   c. A substance that, according to the Brønsted-Lowry Acid-Base Theory, is a proton-donor
   d. A substance that donates its proton and forms its conjugate base in a neutralization reaction

35. Which of the following touch receptors respond to light touch and slower vibrations?
   a. Merkel's discs
   b. Pacinian corpuscles
   c. Meissner's corpuscles
   d. Ruffini endings

36. Nociceptors detect which of the following?
   a. Deep pressure
   b. Vibration
   c. Pain
   d. Temperature

37. A cluster of capillaries that functions as the main filter of the blood entering the kidney is known as which of the following?
   a. The Bowman's capsule
   b. The Loop of Henle
   c. The glomerulus
   d. The nephron

38. What is an alteration in the normal gene sequence called?
   a. DNA mutation
   b. Gene migration
   c. Polygenetic inheritance
   d. Incomplete dominance

39. Blood type is a trait determined by multiple alleles, and two of them are co-dominant: I^A codes for A blood and I^B codes for B blood. i codes for O blood and is recessive to both. If an A heterozygote individual and an O individual have a child, what is the probably that the child will have A blood?
   a. 25%
   b. 50%
   c. 75%
   d. 100%

40. Which of the choices below are the reproductive cells produced by meiosis?
   a. Genes
   b. Alleles
   c. Chromatids
   d. Gametes

41. What is the process of cell division in somatic (most body) cells called?
   a. Mitosis
   b. Meiosis
   c. Respiration
   d. Cytogenesis

42. When human cells divide by meiosis, how many chromosomes do the resulting cells contain?
   a. 96
   b. 54
   c. 46
   d. 23

43. What is the name for the sac-shaped structures in which carbon dioxide and oxygen exchange takes place?
a. Kidneys
b. Medulla oblongata
c. Alveoli
d. Bronchioles

44. What is the pathway of oxygenated blood from the lungs?
a. Lungs to the left atrium, through the mitral valve into the left ventricle, pumped into the aorta upon contraction, then dispersed to tissues via a network of arteries and capillaries
b. Lungs to the right atrium, through the mitral valve into the right ventricle, pumped into the aorta upon contraction, then dispersed to tissues via a network of arteries and veins
c. Lungs to the left atrium, directly to the right aorta, then dispersed to tissues via a network of arteries and capillaries
d. Lungs to the left atrium, through the septum valve, stored in the left ventricles, then dispersed to tissues via a network of arteries and capillaries

45. Which structure serves as the electrical stimulator of the cardiac muscle?
a. The sinoatrial node
b. The left ventricle
c. The aorta
d. The tricuspid valve

46. The primary function of the endocrine system is to maintain which of the following?
a. Heartbeat
b. Respiration
c. Electrolyte and water balance
d. Homeostasis

47. Which of the following organs functions both as an endocrine and exocrine gland?
a. The kidney
b. The spleen
c. The pancreas
d. The stomach

48. How many grams of solid $CaCO_3$ are needed to make 600 mL of a 0.35 M solution? The atomic masses for the elements are as follows: Ca = 40.07 g/mol; C = 12.01 g/mol; O = 15.99 g/mol.
a. 18.3 g
b. 19.7 g
c. 21.0 g
d. 24.2 g

49. How many mL (to the appropriate number of significant figures) of a 12.0 M stock solution of HCl should be added to water to create 250 mL of a 1.50 M solution of HCl?
a. 31.25 mL
b. 32 mL
c. 30 mL
d. 31.3 mL

50. With which genotype would the recessive phenotype appear, if the dominant allele is marked with "A" and the recessive allele is marked with "a"?
  a. AA
  b. aa
  c. Aa
  d. aA

51. What is the name for the structural and functional units that create and transport urine, located in the cortex of the kidney?
  a. Glomerulus
  b. Nephrons
  c. Loops of Henle
  d. Distal tubules

52. The muscular tube that connects the outer surface to the cervix in a woman's birth canal is referred to as which of the following?
  a. The uterus
  b. The cervix
  c. The vagina
  d. The ovaries

53. Which of the following hormones is primarily responsible for regulating metabolism?
  a. Insulin
  b. Testosterone
  c. Adrenaline
  d. Thyroid hormone

Answer Explanations #2

1. C: A hypothesis is a statement that makes a prediction between two variables. The two variables here are the amount of poison and how quickly the rat dies. Choice *C* states that the more poison a rat is given, the more quickly it will die, which is a prediction. Choice *A* is incorrect because it's simply an observation. Choice *B* is incorrect because it's a question posed by the observation but makes no predictions. Choice *D* is incorrect because it's simply a fact.

2. D: Mechanical digestion is physical digestion of food and tearing it into smaller pieces using force. This occurs in the stomach and mouth. Chemical digestion involves chemically changing the food and breaking it down into small organic compounds that can be utilized by the cell to build molecules. The salivary glands in the mouth secrete amylase that breaks down starch, which begins chemical digestion. The stomach contains enzymes such as pepsinogen/pepsin and gastric lipase, which chemically digest protein and fats, respectively. The small intestine continues to digest protein using the enzymes trypsin and chymotrypsin. It also digests fats with the help of bile from the liver and lipase from the pancreas. These organs act as exocrine glands because they secrete substances through a duct. Carbohydrates are digested in the small intestine with the help of pancreatic amylase, gut bacterial flora and fauna, and brush border enzymes like lactose. Brush border enzymes are contained in the towel-like microvilli in the small intestine that soak up nutrients.

3. B: Models are representations of concepts that are impossible to experience directly, such as the 3D representation of DNA, so Choice *B* is correct. Choice *A* is incorrect because theories simply explain why things happen. Choice *C* is incorrect because laws describe how things happen. Choice *D* is false because an observation analyzes situations using human senses.

4. A: A control is the component or group of the experimental design that isn't manipulated—it's the standard against which the resultant findings are compared, so Choice *A* is correct. A variable is an element of the experiment that is able to be manipulated, making Choice *B* false. A constant is a condition of the experiment outside of the hypothesis that remains unchanged in order to isolate the changes in the variables; therefore, Choice *C* is incorrect. Choice *D* is false because collected data are simply recordings of the observed phenomena that result from the experiment.

5. C: An inference is a logical prediction of a why an event occurred based on previous experiences or education. The person in this example knows that plants need water to survive; therefore, the prediction that someone forgot to water the plant is a reasonable inference, hence Choice *C* is correct. A classification is the grouping of events or objects into categories, so Choice *A* is false. An observation analyzes situations using human senses, so Choice *B* is false. Choice *D* is incorrect because collecting is the act of gathering data for analysis.

6. A: The decimal point for this value is located after the final zero. Because the decimal is moved 12 places to the left in order to get it between the *8* and the *6*, then the resulting exponent is positive, so Choice *A* is the correct answer. Choice *B* is false because the decimal has been moved in the wrong direction. Choice *C* is incorrect because the decimal has been moved an incorrect number of times. Choice *D* is false because this value is written to three significant figures, not two.

7. B: The set of results is close to the actual value of the acceleration due to gravity, making the results accurate. However, there is a different value recorded every time, so the results aren't precise, which makes Choice *B* the correct answer.

8. B: To solve this, the number of moles of NaCl needs to be calculated:

First, to find the mass of NaCl, the mass of each of the molecule's atoms is added together as follows:

$$23.0\text{g (Na)} + 35.5\text{g (Cl)} = 58.8\text{g NaCl}$$

Next, the given mass of the substance is multiplied by one mole per total mass of the substance:

$$4.0\text{g NaCl} \times (1 \text{ mol NaCl}/58.5\text{g NaCl}) = 0.068 \text{ mol NaCl}$$

Finally, the moles are divided by the number of liters of the solution to find the molarity:

$$(0.068 \text{ mol NaCl})/(0.120\text{L}) = 0.57 \text{ M NaCl}$$

Choice *A* incorporates a miscalculation for the molar mass of NaCl, and Choices *C* and *D* both incorporate a miscalculation by not converting mL into liters (L), so they are incorrect by a factor of 10.

9. C: According to the *ideal gas law* ($PV = nRT$), if volume is constant, the temperature is directly related to the pressure in a system. Therefore, if the pressure increases, the temperature will increase in direct proportion. Choice *A* would not be possible, since the system is closed and a change is occurring, so the temperature will change. Choice *B* incorrectly exhibits an inverse relationship between pressure and temperature, or $P = 1/T$. Choice *D* is incorrect because even without actual values for the variables, the relationship and proportions can be determined.

10. B: A solid produced during a reaction is called a *precipitate.* In a neutralization reaction, the products (an acid and a base) react to form a salt and water. Choice *A*, an oxidation reaction, involves the transfer of an electron. Choice *C*, a synthesis reaction, involves the joining of two molecules to form a single molecule. Choice *D*, a decomposition reaction, involves the separation of a molecule into two other molecules.

11. C: Molecules that are soluble in lipids, like fats, sterols, and vitamins (A, D, E and K), for example, are able to move in and out of a cell using passive transport. Water and oxygen are also able to move in and out of the cell without the use of cellular energy. Complex sugars and non-lipid soluble molecules are too large to move through the cell membrane without relying on active transport mechanisms. Molecules naturally move from areas of high concentration to those of lower concentration. It requires active transport to move molecules in the opposite direction, as suggested by Choice *D*.

12. A: Both types of cells are enclosed by a cell membrane, which is selectively permeable. Selective permeability means essentially that it is a gatekeeper, allowing certain molecules and ions in and out, and keeping unwanted ones at bay, at least until they are ready for use. Prokaryotes contain a nucleoid and do not have organelles; eukaryotes contain a nucleus enclosed by a membrane, as well as organelles.

13. B: During telophase, two nuclei form at each end of the cell and nuclear envelopes begin to form around each nucleus. The nucleoli reappear, and the chromosomes become less compact. The microtubules are broken down by the cell, and mitosis is complete. The process begins with prophase as the mitotic spindles begin to form from centrosomes. Prometaphase follows, with the breakdown of the nuclear envelope and the further condensing of the chromosomes. Next, metaphase occurs when the microtubules are stretched across the cell and the chromosomes align at the metaphase plate. Finally, in the last step before telophase, anaphase occurs as the sister chromatids break apart and form chromosomes.

14. D: Meiosis has the same phases as mitosis, except that they occur twice—once in meiosis I and once in meiosis II. During meiosis I, the cell splits into two. Each cell contains two sets of chromosomes. Next, during meiosis II, the two intermediate daughter cells divide again, producing four total haploid cells that each contain one set of chromosomes.

15. B: The secondary structure of a protein refers to the folds and coils that are formed by hydrogen bonding between the slightly charged atoms of the polypeptide backbone. The primary structure is the sequence of amino acids, similar to the letters in a long word. The tertiary structure is the overall shape of the molecule that results from the interactions between the side chains that are linked to the polypeptide backbone. The quaternary structure is the complete protein structure that occurs when a protein is made up of two or more polypeptide chains.

16. A: The sensory threshold is the smallest amount of stimulus that is required for an individual to experience one of the senses. For example, during a hearing test, the sensory threshold would be the quietest sound that a person could detect. This threshold is an important indicator of whether a person's senses are working within a normal range.

17. C: Generally, all sensory pathways that extend from the sensory receptor to the brain are composed of three long neurons called the primary, secondary, and tertiary neurons. The primary one stretches from the sensory receptor to the dorsal root ganglion of the spinal nerve; the secondary one stretches from the cell body of the primary neuron to the spinal cord or the brain stem; the tertiary one stretches from the cell body of the secondary one into the thalamus. Each type of sense, such as touch, hearing, and vision, has a different pathway designed specifically for that sensation.

18. C: Visual processing begins in the retina. When an individual sees an image, it is taken in through the cornea and lens and then transmitted upside down onto the retina. The cells in the retina process what is being seen and then send signals to the ganglion cells, whose axons make up the optic nerve. The optic nerve cells connect the retina to the brain, which is where the processing of the visual information is completed and the images are returned to their proper orientation.

19. D: Arteries carry oxygen-rich blood from the heart to the other tissues of the body. Veins carry oxygen-poor blood back to the heart. Intestines carry digested food through the body. Bronchioles are passageways that carry air from the nose and mouth to the lungs.

20. C: The somatic nervous system is the voluntary nervous system, responsible for voluntary movement. It includes nerves that transmit signals from the brain to the muscles of the body. Breathing is controlled by the autonomic nervous system. Thought and fear are complex processes that occur in the brain, which is part of the central nervous system.

21. D: Catabolism is the process of breaking large molecules into smaller molecules to release energy for work. Carbohydrates and fats are catabolized to provide energy for exercise and daily activities. Anabolism synthesizes larger molecules from smaller constituent building blocks. Bioenergetics and metabolism are more general terms involving overall energy production and usage.

22. A: Platelets are the blood components responsible for clotting. There are between 150,000 and 450,000 platelets in healthy blood. When a clot forms, platelets adhere to the injured area of the vessel and promote a molecular cascade that results in adherence of more platelets. Ultimately, the platelet aggregation results in recruitment of a protein called fibrin, which adds structure to the clot. Too many platelets can cause clotting disorders. Not enough leads to bleeding disorders.

23. A: Choice B might be an attractive answer choice, but neutrophils are part of the innate immune system and are not considered part of the primary immune response. The first event that happens in a primary immune response is that macrophages ingest pathogens and display their antigens. Then, they secrete interleukin 1 to recruit helper T cells. Once helper T cells are activated, they secrete interleukin 2 to simulate plasma B and killer T cell production. Only then can plasma B make the pathogen specific antibodies.

24. C: The epididymis stores sperm and is a coiled tube located near the testes. The immature sperm that enters the epididymis from the testes migrates through the 20-foot long epididymis tube in about two weeks, where viable sperm are concentrated at the end. The vas deferens is a tube that transports mature sperm from the epididymis to the urethra. Seminal vesicles are pouches attached that add fructose to the ejaculate to provide energy for sperm. The prostate gland excretes fluid that makes up about a third of semen released during ejaculation. The fluid reduces semen viscosity and contains enzymes that aid in sperm functioning; both effects increase sperm motility and ultimate success.

25. A: All three joint types given are synovial joints, allowing for a fair amount of movement (compared with fibrous and cartilaginous joints). Of the three given, hinge joints, such as the elbow, permit the least motion because they are uniaxial and permit movement in only one plane. Saddle joints and condyloid joints both have reciprocating surfaces that mate with one another and allow a variety of motions in numerous planes, but saddle joints, such as the thumb carpal metacarpal joint allow more motion than condyloid joints. In saddle joints, two concave surfaces articulate, and in a condyloid joint, such as the wrist, a concave surface articulates with a convex surface, allowing motion in mainly two planes.

26. D: All of the above. Synergists are responsible for helping the primary movers or agonists carry out their specific movements. Synergists can help stabilize the point of origin of a muscle or provide extra pull near the insertion; in this sense, they can be considered "fixators." Some synergists can help prevent undesired movement at a joint. For example, for elbow flexion, the brachialis is a synergist to the biceps.

27. D: Bony matrix is an intricate lattice of collagen fibers and mineral salts, particularly calcium and phosphorus. The mineral salts are strong but brittle, and the collagen fibers are weak but flexible, so the combination of the two makes bone resistant to shattering and able to withstand the normal forces applied to it.

28. D: The natural tendency in the universe is to increase entropy and disorder, so organisms undergo biochemical processes and use free energy and matter to stabilize their internal environment against this changing external environment in a process called homeostasis. Organisms strive to maintain physiologic factors such as body temperature, blood pH, and fluid balance in equilibrium or around a set point. Choice *A* is incorrect because homeostasis in organisms involves trying to maintain internal conditions around their set point, which doesn't always involve increasing body temperature, blood pH, and fluid balance; the organism may need to work to decrease these values. Choice *B* is incorrect because maintaining homeostasis expends energy; it does not absorb it. Choice *C* is incorrect because homeostasis involves maintaining order in the organism's internal, not external, environment.

29. A: All three types of muscle tissue (skeletal, cardiac, and smooth) share four important properties: They are contractile, meaning they can shorten and pull on connective tissue; excitable, meaning they respond to stimuli; elastic, meaning they rebound to their original length after a contraction; and extensible, meaning they can be stretched repeatedly, but maintain the ability to contract. While skeletal muscle is under voluntary control, both cardiac and smooth muscle are involuntary.

30. D: The autonomic nervous system (ANS), a division of the PNS controls involuntary functions such as breathing, heart rate, blood pressure, digestion, and body temperature via the antagonistic parasympathetic and sympathetic nerves. Choices *A, B,* and *C* are incorrect because they describe characteristics of the somatic nervous system, which is the other division of the PNS. It controls skeletal muscles via afferent and efferent neurons. Afferent neurons carry sensory messages from skeletal muscles, skin, or sensory organs to the CNS, while efferent neurons relay motor messages from the CNS to skeletal muscles, skin, or sensory organs. While skeletal muscle movement is under voluntary control, the somatic nervous system also plays a role in the involuntary reflex arc.

31. C: The sinuses function to warm, filter, and humidify air that is inhaled. Choice *A* is incorrect because mucus traps airborne pathogens. Choice *B* is incorrect because the epiglottis is the structure in the pharynx that covers the trachea during swallowing to prevent food from entering it. Lastly, Choice *D,* sweeping away pathogens and directing them toward the top of the trachea, is the function of cilia. Respiratory structures, such as the nasal passages and trachea, are lined with mucus and cilia.

32. A: The renal pelvis acts like funnel by delivering the urine from the millions of the collecting tubules to the ureters. It is the most central part of the kidney. The renal cortex is the outer layer of the kidney, while the renal medulla is the inner layer. The renal medulla contains the functional units of the kidneys—nephrons—which function to filter the blood. Choice *D,* Bowman's capsule, is the name for the structure that covers the glomeruli.

33. B: The pelvic cavity is the area formed by the bones of the hip. Housed in that space are the urinary bladder, urethra, ureters, anus, and rectum. This hollow space also contains the uterus in females. The other body cavities listed contain other organs not specified in the question.

34. B: An amphoteric substance can act as an acid or a base depending on the properties of the solute. Water is a common example and because of its amphoteric property, it serves as a universal solvent. Choice *A* is incorrect because it describes electrolytes. Choices *C* and *D* are incorrect because they both describe an acid.

35. C: The body has a variety of touch receptors that detect tactile information, which gets carried via general somatic afferents and general visceral afferents to the CNS. Meissner's corpuscles respond to light touch and slower vibrations. Choice *A* is incorrect because Merkel's discs respond to sustained pressure. Choice *B* is incorrect because Pacinian corpuscles detect rapid vibration and in the skin and fascia, and Choice *D* is incorrect because Ruffini endings detect deep touch and tension in the skin and fascia.

36. C: Nociceptor are a type of touch receptor that detect pain. They relay this information to the CNS via afferent nerves. Efferent nerves then carry motor signals back, which may cause muscle contractions that move the body part away from the source of the pain. For example, if a person places his or her hand on a hot stove, nociceptors detect the painful burning sensation. Then they relay this information up to the brain, where a response signal travels down the efferent nerves that control the muscles of the shoulder, arm, forearm, and/or hand to retract the hand from the hot stove. Choice *A* is incorrect because that is mainly the function of Ruffini endings. Choice *B* describes the function of several mechanoreceptors such as Meissner's corpuscles, Pacinian corpuscles, and Merkel's discs. Lastly, thermoreceptors detect changes in pressure, so Choice *D* is incorrect.

37. C: A cluster of capillaries that functions as the main filter of the blood entering the kidney is known as the glomerulus, so Choice *C* is correct. The Bowman's capsule surrounds the glomerulus and receives fluid and solutes from it; therefore, Choice *A* is incorrect. The loop of Henle is a part of the kidney tubule where water and nutrients are reabsorbed, so *B* is false. The nephron is the unit containing all of these anatomical features, making Choice *D* incorrect as well.

38. A: An alteration in the normal gene sequence is called a DNA point mutation. Mutations can be harmful, neutral, or even beneficial. Sometimes, as seen in natural selection, a genetic mutation can improve fitness, providing an adaptation that will aid in survival. DNA mutations can happen as a result of environmental damage, for example, from radiation or chemicals. Mutations can also happen during cell replication, as a result of incorrect pairing of complementary nucleotides by DNA polymerase. There are also chromosomal mutations as well, where entire segments of chromosomes can be deleted, inverted, duplicated, or sent or received from a different chromosome.

39. B: 50%. According to the Punnett square, the child has a 2 out of 4 chance of having A-type blood, since the dominant allele I^A is present in two of the four possible offspring. The O-type blood allele is masked by the A-type blood allele since it is recessive.

I^A i	ii
I^A i	ii

40. D: Reproductive cells are referred to as gametes: egg (female) and sperm (male). These cells have only 1 set of 23 chromosomes and are haploid so that when they combine during fertilization, the zygote has the correct diploid number, 46. Reproductive cell division is called meiosis, which is different from mitosis, the type of division process for body (somatic) cells.

41. A: The process of cell division in somatic is mitosis. In interphase, which precedes mitosis, cells prepare for division by copying their DNA. Once mitotic machinery has been assembled in interphase, mitosis occurs, which has four distinct phases: prophase, metaphase, anaphase, and telophase, followed by cytokinesis, which is the final splitting of the cytoplasm. The two diploid daughter cells are genetically identical to the parent cell.

42. D: Human gametes each contain 23 chromosomes. This is referred to as haploid—half the number of the original germ cell (46). Germ cells are diploid precursors of the haploid egg and sperm. Meiosis has two major phases, each of which is characterized by sub-phases similar to mitosis. In Meiosis I, the DNA of the parent cell is duplicated in interphase, just like in mitosis. Starting with prophase I, things become a little different. Two homologous chromosomes form a tetrad, cross over, and exchange genetic content. Each shuffled chromosome of the tetrad migrates to the cell's poles, and two haploid daughter cells are formed. In Meiosis II, each daughter undergoes another division more similar to mitosis (with the exception of the fact that there is no interphase), resulting in four genetically-different cells, each with only ½ of the chromosomal material of the original germ cell.

43. C: The alveoli are small sac-shaped structures at the end of the bronchioles where gas exchange takes place. The bronchioles are tubes through which air travels. The kidneys and medulla oblongata do not directly affect oxygen and carbon dioxide exchange.

44. A: This path correctly describes where blood flows after leaving the lungs in order to reach the tissues of the body. The other paths listed are incorrect, out of order, and list fictitious structures.

45. A: This node is the primary stimulator of electrical activity in the heart. The other structures listed play a role in blood flow, but do not deal with electrical stimulation.

46. D: The primary function of the endocrine system is to maintain homeostasis, which means it makes constant adjustments to the body's systemic physiology to maintain a stable internal environment. Homeostasis requires an adequate heart rate, respirations, and water and electrolyte balance, as well as many other physiological processes; therefore, Choice *D* is the correct answer. The other answers are false because they represent only one aspect of homeostasis.

47. C: The pancreas functions as an exocrine gland because it secretes enzymes that break down food components. It also functions as an endocrine gland because it secretes hormones that regulate blood sugar levels. The kidney isn't a gland; it is an organ located directly below the adrenal glands. The stomach is an exocrine gland because it secretes hydrochloric acid. Like the kidney, the spleen is an organ, not a gland. Therefore, the only correct answer is *C*.

48. C: To make a solution from a pure solid, the total molecular weight of the substance must be calculated and then the proper mass of the substance in grams must be added to water to make a solution. To calculate the total molecular weight, the individual molecular weights must be added. Finally, the mass of substance needed to make the solution can be calculated.

$$600mL\ CaCO_3 \times \frac{0.350\ m * l}{1000mL} \times \frac{100g\ CaCO_3}{1mol\ CaCO_3} = 21g\ CaCO_3$$

The calculations reveal that Choice *C* is the correct answer. All other reported values are incorrect.

49. D: Preparing a solution from a stock is simply a process of dilution by adding water to a certain amount of the stock. The amount of stock to use can be calculated using a formula and algebra:

$$V_S = \frac{M_D V_D}{M_S}$$

$$M_D = 1.5$$

$$V_D = 250ml$$

$$M_S = 12.0M$$

$$V_S = \frac{(1.5M)(250ml)}{12.0M} = 31.3ml$$

Because the given values are written to three significant figures, the answer should also be written in three significant figures, making Choice *D* the correct answer. The other answer choices are either incorrect values or reported to an incorrect number of significant figures.

50. B: Dominant alleles are considered to have stronger phenotypes and, when mixed with recessive alleles, will mask the recessive trait. The recessive trait would only appear as the phenotype when the allele combination is "aa" because a dominant allele is not present to mask it.

51. B: The nephron is the structural and functional unit that creates and transports urine. The glomerulus, loop of Henle, and distal tubule are parts of the nephron involved in creating and transporting urine depending on their respective functions, so Choice *B* is the correct answer. Choices *A*, *C*, and *D* are incorrect, as they represent only a part of the unit.

52. C: The uterus and ovaries aren't part of the birth canal, so Choices *A* and *D* are false. The cervix is the uppermost portion of the birth canal, so Choice *B* is incorrect, making Choice *C* the correct answer, as the vagina is the muscular tube on the lowermost portion of the birth canal that connects the exterior environment to the cervix.

53. D: Thyroid hormone is responsible for regulating metabolism, so Choice *D* is the correct answer. Insulin is involved in glucose uptake into tissues, testosterone is for sperm production and secondary male sexual characteristics, and adrenaline is responsible for mechanisms in the flight-or-fight response, making Choices *A*, *B*, and *C* incorrect.

Dear ATI TEAS Test Taker,

We would like to start by thanking you for purchasing this practice test book for your ATI TEAS exam. We hope that we exceeded your expectations.

We strive to make our practice questions as similar as possible to what you will encounter on test day. With that being said, if you found something that you feel was not up to your standards, please send us an email and let us know.

We would also like to let you know about other books in our catalog that may interest you.

HESI

This can be found on Amazon: amazon.com/dp/1628456019

CEN

amazon.com/dp/1628454768

We have study guides in a wide variety of fields. If the one you are looking for isn't listed above, then try searching for it on Amazon or send us an email.

Thanks Again and Happy Testing!
Product Development Team
info@studyguideteam.com

Interested in buying more than 10 copies of our product? Contact us about bulk discounts:

bulkorders@studyguideteam.com

FREE Test Taking Tips DVD Offer

To help us better serve you, we have developed a Test Taking Tips DVD that we would like to give you for FREE. **This DVD covers world-class test taking tips that you can use to be even more successful when you are taking your test.**

All that we ask is that you email us your feedback about your study guide. Please let us know what you thought about it – whether that is good, bad or indifferent.

To get your **FREE Test Taking Tips DVD**, email freedvd@studyguideteam.com with "FREE DVD" in the subject line and the following information in the body of the email:

a. The title of your study guide.

b. Your product rating on a scale of 1-5, with 5 being the highest rating.

c. Your feedback about the study guide. What did you think of it?

d. Your full name and shipping address to send your free DVD.

If you have any questions or concerns, please don't hesitate to contact us at freedvd@studyguideteam.com.

Thanks again!

37471007R00098

Made in the USA
Lexington, KY
26 April 2019